WOMEN
IN COMBAT

WOMEN
IN COMBAT
FEMINISM GOES TO WAR

MARK C.ATKINS

Foreword By MAJOR GENERAL PATRICK HENRY BRADY, USA (Ret)
Recipient of the Congressional Medal of Honor

SHOTWELL PUBLISHING
Columbia, South Carolina

Produced in the REPUBLIC OF SOUTH CAROLINA by

SHOTWELL PUBLISHING LLC
Post Office Box 2592
Columbia, So. Carolina 29202

ShotwellPublishing.com

Cover Illustration: Jeffery Forrest
Cover Design: Hazel's Dream – Boo Jackson TCB

Second Edition

ISBN-13: 978-1-947660-24-3
ISBN-10: 1-947660-24-1

10 9 8 7 6 5 4 3 2 1

PRE-PUBLICATION COMMENTS ON
WOMEN IN COMBAT

Mark Atkins has written an excellent book on a key national security issue which affects the capability of our military forces - Women in Combat. *He thoroughly discusses the pros and cons. It is clear there are many viable roles for women in our military, but combat is not one of them!*

—Admiral James "Ace" Lyons, USN Retired, Commander US Pacific Fleet 1985-1987, two time recipient of the Distinguished Service Medal

♦ ♦ ♦

Women in Combat *is a great, educational read! .. ... He rightly blames the feminist movement and I will blame the feminization of our military at least in part for its low morale today. I recommend that every American who severes our military and understands its purpose read this book and share and discuss it with others.*

— Colonel Jim Harding, USAF Retired, recipient of the Air Force Cross and the 25th most decorated American fighting man of all time

♦ ♦ ♦

Congratulations to Mark Atkins for his plain-talk grappling with the effects of nature and 'isms' that swirl in a controversial yet still-developing "woman-warrior" phenomenon that has ominous, metaphysical proportions. Man's indifferent determination to regularize women in military armed forces as though they are men is a unique signal of ongoing destructive transformation in our civilization as once informed, if imperfectly, by western Christendom . . . . Atkins contends colorfully with the gamut of controversial audacity, and leads readers to face reality, the divinely-created origination of all that is, in particular of Man as male-and-female. That truth, and a "fear of God", have largely dissipated from public consensus and conviction. This stirring book comes at a particularly timely moment, as a Congressionally-appointed National Commission begins considering recommendations for the future of Military Selective Service, with perhaps the draft registration of women. See inspire2serve.gov. For potential recruits and their parents, to policy-makers and church pastors, this fast-moving collection of Atkins' critical reflections and colorful vignettes depict a situation deserving sober and decisive attention.

—Robert H Miller, CAPT, USN Retired
Executive Director, Hope For America

♦ ♦ ♦

In his book, Women in Combat, *Mark Atkins pulls the reins in on the progressive movement's rush to place women in unlimited roles within the U.S. military. The political hacks and policy wonks still in uniform at the time rushed right over the question of whether women should serve in combat arms units — where historically hand-to-hand combat was the norm — and jumped straightway into a debate on whether they can, and whether there is any limit to where they could serve. Mark takes us back to the question of whether or not they should, and his splendid polemic explains why culturally, socially, historically and practically they should not. It is an excellent thought piece and he is right; it will cause feminists to blow their last gasket.*

—*Tony Foster, USMA graduate, recipient of the General Douglas MacArthur Leadership Award*

♦ ♦ ♦

Mark Atkins, a successful businessman and self-described 'country philosopher,' has written a simple, clear, brief, but forceful treatise against both the fact of women in the armed forces being assigned to military combat (and soon perhaps drafted into the services) and the false 'philosophy' and pseudo-logic behind these human outrages. Though Atkins insists that he is not a 'professional' writer, he modestly overlooks the fact that one need not be an author by trade to write effectively, as many a literate man (and woman) of common sense abundantly proved in less specialized eras than today. Among the many good uses to which I foresee Women in Combat *being put is its parental insertion into the hands of girls and young women misguidedly contemplating either a stint or career in the military.*

—*Chilton Williamson, Jr., author, novelist, and editor of* Chronicles: A Magazine of American Culture

♦ ♦ ♦

George Orwell said about our times: 'We have now sunk to a depth at which restatement of the obvious is the first duty of intelligent men.' Mark Atkins is one of those men who has ably restated the truth: that the idea of women in combat is a violation of human nature and Western civilization. Atkins' book presents the conclusive case."

—*Clyde N. Wilson, Emeritus Distinguished Professor of History, University of South Carolina*

♦ ♦ ♦

Today the Will supersedes God and Nature. It is against the law to display the Ten Commandments in State Court Houses. Man and Woman are abolished in favor of the arbitrary concept of 'gender.' A man can become a woman, and a woman a man. Same sex couples can become 'man and wife.' Women in Combat *exposes this rebellion in the military with arguments unanswerable to any mind not breathing the fumes of political correctness.*

—*Donald W. Livingston, Professor of Philosophy, Emory University; Fellow, Institute for Advanced Studies in the Humanities, University of Edinburgh*

♦ ♦ ♦

Informally written, the book is suitable for a wide audience and succinctly distills major themes on this important issue into easily understandable concepts. Drawing on history, psychology, biology and philosophy, Atkins outlines the case against wide scale integration of women into combat units, and accurately describes the motives of those who are pushing for that change. In his view, it is Feminism running amok that is the primary advocate of that absurd policy, and the author does a solid job of making his case that feminism should be vying for the title of public enemy number one on any conservative Americans' hit parade. Given that he wears his Christianity on his sleeve in the book, Mark Atkins is bound to be attacked as a zealot, a misogynist, and a homophobe. Frankly, when that happens, it will be doing more to expose his opponents' character flaws than his own. The author is an old fashioned, patriotic Christian American and making no apologies for it. People like him built America into the world's first hyper power. It will be interesting to see if America still has enough men and women like him to save it from the fate that has eventually befallen every other major economic and military superpower on Earth.

—James B White, Capt., US Army, USMA Class of 1980

♦ ♦ ♦

In his book Women in Combat, *Mark Atkins provides a startling reality check on the rush to succumb to political correctness- no matter the cost. His provocative and entertaining writing style engages the reader's mind to think about the issues of women in combat from different perspectives. Whether one agrees with his positions or not, the book will crystalize the important components of this critical societal issue. Moreover, this work delves into related issues in our society and is extremely thought provoking.*

—Rich Gregg, Major USMC (Retired)

♦ ♦ ♦

Though never having served in the military I did find it unnerving when we began to allow women into combat situations. The logic against institutionalizing women in combat is shockingly simple yet somehow has been hiding in plain sight. Clearly, we should not do this. More importantly we are taking the woman out of the position assigned to her by God as revealed in His Word. When we are out of God's will we will always find ourselves in a struggle. Atkins' declaration of the results of rebelling against His nature says it all: our healthy-nature produces strength, our unhealthy-nature produces weakness. Institutionalizing women in combat can only produce weakness.

—Roger Anghis, Columnist, Radio Host

♦ ♦ ♦

Mark Atkins book, Women in Combat, *is a thoughtful, logical presentation of the obvious. The snowflakes are going to melt if they dare to read this concise unveiling of the harm feminism has done and continues to do to our people. Atkins lays out a powerful argument for not expanding the role of women in the military to combat specialties. Being in the first class at West Point with women, I can attest to how the standards were lowered to ensure they were 'successful' and graduated. As Atkins points out, our enemies will not be so accommodating. I spent twenty-four years in the Army watching the struggle to deal with the unnatural presence of women dictated by politicians whose agenda was certainly not national security. I can only grimace at the negative impact openly gay and transgender 'soldiers' are having on readiness.*

—Randy E. Geiger,LTC, U.S. Army (Retired), West Point class of 1980

♦ ♦ ♦

Trump's Secretary of Defense—should seriously read Women in Combat, *by Tennessee writer, Mark Atkins. The whole country could benefit if a friendly fellow flag officer brings Mattis this book. Atkins deals with all of the unthinkable, unspeakable militant-feminist PC matters self-evident in the title, as illustrated here in the writer's own words: Combat "is not for boys. It is not for old men. It is not for video-game-warrior-betamales. It is certainly not for women. They could not endure it in the days of swords and horses, and they cannot endure it today in an age of rifles and Humvees." Enough said.*

—Marcus R. Hayes, Ph.D., LTC, U.S. Army (Retired)

♦ ♦ ♦

In Women in Combat, *Mark Atkins highlights the unnatural, foolish and immoral agenda of placing women in combat. Identifying this absurdity as a recent historical development that began in the 20th Century feminist movement, he reasons with a sober poignancy as to the upheaval it has brought to the Military and to our culture. I might add . . . women were never numbered with Troops in the Bible. The real issue has never been can she fight, but should she? The Biblical answer is No! What Mark Atkins writes makes sense because this is still a God-ordered universe." " . . .Woman was not made for this, O man, to be prostituted as common. O you subverters of all decency, who use men, as if they were women, and lead out women to war, as if they were men! This is the work of the devil, to subvert and confound all things, to overleap the boundaries that have been appointed from the beginning and remove those which God has set to nature. . . . " (John Chrysostom (c. 347-407), Archbishop of Constantinople, Homily 5 on Titus)*

—Steve Brown Capt., CHC, USN (Retired) President, Associated Gospel Churches

♦ ♦ ♦

Mark Atkins has written a book that never should have been written—and not because he is killing trees, wasting the reader's time, or worst of all, deceiving us. Rather, Women in Combat *only could have been written in an age of thoroughgoing cultural collapse. There yet remain a handful of eccentrics—I am one of them—who believe that the sex designed by God to give and nurture human life ought not be sent forth to take human life, much less in the horrific and immoral ways of the modern battlefield. In our age of executing our own children in the womb, however, sending our mothers, wives, sisters, and daughters off to kill seems bizarrely fitting. Of course, in the light of the natural law, to say nothing of the divine law, a woman's bearing arms in aggression is nothing less than a deliberate mixture of confusion and wickedness. If Atkins wakes up one soul to this moral truth, all of the time, effort, and money spent on this book will have been a bargain.*

—Christopher Check, President Catholic Answers, Captain, USMC, 1987 - 1994, Desert Shield

♦ ♦ ♦

Mark Atkins' superb, timely book, Women in Combat, *is a must-read for today. Our country, or "people" as Mark explains, face a major crisis, a spiritual, intellectual crisis where simple, natural, vital truth is being put to flight by a radical, unnatural, suicidal falsehood. It is a religious, philosophical conflict which is even now creating a military crisis that will surely result in a national catastrophe from which we may never recover, unless we, as a people, heed his warning against the unnatural insanity of institutionalizing women in combat. Mark's book is a must-read for legislators whose constitutional duty is to resolutely stand against the radical political feminist agenda demanding "equality of the sexes" in our nation's military forces and to legislate for natural, rational laws which govern that military. It is a must-read for the combat proven senior commanders of our people's war fighters who must now muster the moral courage to defy the "political correctness" of the day and speak the truth to power: putting women in combat is a militarily ludicrous idea. And it is a must-read for God's ministers who are called to 'walk point' in spiritual and moral combat. First, women cannot fight, that is a fact. Second, in combat, men get killed trying to protect the women. And third, when the enemy finds that he is facing a unit with women combatants, he will not retreat, he will not surrender, he will fight to the death rather than be beaten by women. Mark Atkins has explained why this is true and always will be true. May God restore sanity to our people's leaders before it is too late.*

—Captain James F. Poe, CHC, USN (Ret), Chairman, International Conference of Evangelical Chaplain Endorsers

... glorious Hektor held out his arms to his baby,
who shrank back to his fair-girdled nurse's bosom
screaming, and frightened at the aspect of his own father,
terrified as he saw the bronze and the crest with its horse-hair,
nodding dreadfully, as he thought, from the peak of the helmet.
Then his beloved father laughed out, and his honored mother,
and at once glorious Hektor lifted from his head the helmet
and laid it in all its shining upon the ground. Then taking
up his dear son he tossed him about in his arms, and kissed him,
and lifted his voice in prayer to Zeus and the other immortals:
"Zeus, and you other immortals, grant that this boy, who is my son,
may be as I am, pre-eminent among the Trojans,
great in strength, as I am, and rule strongly over Ilion;
and some day let them say of him: 'He is better by far than his father,'
as he comes in from the fighting; and let him kill his enemy
and bring home the blooded spoils, and delight the heart of his mother."
So speaking he set his child again in the arms of his beloved
wife, who took him back again to her fragrant bosom
smiling in her tears; and her husband saw, and took pity on her,
and stroked her with his hand, and called her by name and spoke to her:
"Poor Andromache! Why does your heart sorrow so much for me?
No man is going to hurl me to Hades, unless it is fated,
but as for fate, I think that no man yet has escaped it
once it has taken its first form, neither brave man nor coward.
Go therefore back to our house, and take up your own work,
the loom and the distaff, and see to it that your handmaidens
ply their work also; but the men must see to the fighting,
all men who are the people of Ilion, but I beyond others."

ઠ

From Homer's Iliad, *Book Six, ll. 466 – 494; trans. Richard Lattimore,*
1951. Hector's farewell to his son and wife on the day of his death

CONTENTS

FOREWORD
Patrick Henry Brady, MG, USA (Ret)

THIS BOOK SHOULD BE REQUIRED reading for the current and future military leadership of America. The two absolutely vital elements of America's survival are a free and honest media and a strong military. One keeps us free and the other keeps us secure. Our media is not honest and our military has been beyond hollow. Thankfully, President Trump is addressing military readiness. He has already addressed the folly of transgenders. He needs to address the homosexual issue and and our daughters and wives in foxholes. We will never be as ready as we could until we correct these follies.

After commanding a medical unit, all men, in combat, I commanded a medical battalion, many women, in peacetime. These units are not direct combat units but do spend a lot of time on the battlefield and are exposed to enemy fire and casualties. But nothing like the exposure of the grunts in the mud and grime humping for days and weeks at a time. My rule in the battalion was standards not sex governed. This was during the 70s, a tough time for drugs, discipline and recruiting in the military.

Here is what I found. As a result of competition, my driver and our color guard, highly contested duty, were women. The women had less disciplinary problems than the men. In administrative jobs they were at least equal to men. But most could not carry their load physically – loading liters in choppers, carrying wounded to safety, even lifting tool chests. As a result the men covered for them often causing us to use two people when one should have done the job – all of which effected readiness. They were not effective in the field becoming less functional over even a three day period when issues of hygiene, and feminine hygiene, literally knocked them out and we had to jerry rig showers to keep them going wasting valuable time.

And they got pregnant which took out key jobs at critical times. Pregnant females cannot deploy and some will get pregnant to avoid it.

Other sexual distractions, favoritisms, fraternization, assault are also readiness disruptions and follow women throughout the military and even in our military academies. I had serious problems with wives whose husbands shared standby shacks with women over night. This would go on for weeks in direct combat units, think tank crews. Male bonding, immeasurable to success in combat, would be damaged. All in all, the women posed an insane burden on readiness.

At the time, perhaps due to recruiting issues, expanding to role of females was being considered. My division commander asked my opinion on the issue. I told him I would not want females with me working the battlefield let alone in direct combat. I told him I would not want my daughters in a unit half women going bayonet to bayonet with an enemy unit one hundred percent men. Those comments almost cost me my career because my immediate superior, a rare progressive at the time, disagreed. We have many more like him today.

The move to teach our daughters and mothers to kill is defended using the same criteria I used in my battalion, standards not sex govern. It does not work. Most men will not demand from women as they do other men -- thankfully. And there is no intention to do so despite what we hear. Listen to a former military leader, General Martin Dempsey: "If we decide that a particular standard is so high that a woman couldn't make it, the burden is now on the service to come back and explain to the secretary, why is it that high? Does it really have to be that high?" Those standards have been set over hundreds of years of combat! We should change them to satisfy the crazes of a politician's feminist supporters? Imagine how General George Patton and all the leaders who founded and secured this country would react to those comments.

It is difficult to understand the psychosis behind politicians who would put our wives and daughters in foxholes and the military leadership that would stand still for it. (It is important to know we are talking foxholes, down and dirty combat. If your hypothesis is that future wars will be sterile, pushbutton, cyber warfare, of course women can do that as well as men. But your hypothesis is insane).

I'm reminded of a comment by a former four star officer who said in one breath the military should reflect our society and then another that the military is unique in our society? Imagine a military that reflects our society, that is not unique, in discipline, voting, fitness, morals, yuck! The majority of our young people are not fit physically and otherwise to meet minimal standards to serve in the military. Should we apportion them in according to society, so many obese, so many multi sexual, so many handicapped, so many dropouts, so many criminals. Give me a break.

At the core of this travesty was Barack Obama and Leon Panetta. Both were politicians with zero combat experience, fearful of the feminists and driven by the hope of increased female votes. Both stood down while Americans were massacred in Benghazi. Obama could be excused since he did not know the difference between a corpse and a corps. Panetta, as Secretary of Defense, rationalized that: "(The) basic principle is that you don't deploy forces into harm's way without knowing what's going on; without having some real-time information about what's taking place."

That statement takes its place beside "leading from behind" as two of the most inane statements from any leader in the history of warfare.

For many Americans it is hard to believe that Panetta could top his statement in defense of the administration's tragic bungling of the terrorists' massacre in Benghazi: But he did. In justification of the administration's policy to put women in foxholes he alleged women in [Direct] combat strengthen our military. His statement on Benghazi is a contradiction of every war we have fought and the ethos of every warrior who ever fought in those wars. Panetta's statement extolling the readiness multiplier of women leading bayonet charges is beyond the pale.

Obama was fortunate in the class of military he inherited, and he was around long enough to mold many in his image and likeness. We suffer from Obama holdovers today. Few of the current and recent military leadership has had any serious combat (in the trenches) experience and it was evident in their reaction to the formation of a quad sexual military (at the time, now many more sexes) and women in foxholes decisions. No one resigned. It is hard to believe they did not know the effect on readiness both decisions would have, but they cowered, not a word.

No serous person could believe that women in foxholes and sodomy in the barracks will do anything but reduce readiness. Homosexuals bring with them serious health and deployment issues. Homosexuals cannot give blood and may not be deployable. Every warrior is a walking blood bank. Who would want their son or daughter to receive a blood transfusion from a homosexual – let alone themselves? The NBA stops a basketball game for a drop of blood because of the threat of infection, the Magic Johnson rule, Johnson had AIDs. The battlefield is full of blood. Do we think less of our soldiers than the NBA does of its players? What will be the reaction when a warrior sees his commanding officer dancing and romancing another man at the club – or if he is hit on by a homosexual? Yet we are told these changes will improve readiness

World War II was influenced by combat veterans from World War I. In Korea we had the veterans of World War II, and in Vietnam the combat veterans of both World War II and Korea. The Vietnam veteran won Desert Storm one of the most remarkable victories in military history. All those warriors, and their leadership, are gone and we have seen a military with dismal leadership resulting in unprecedented rates of suicide, PTSD and security breaches. We had one high ranking officer lament that the terrorist's massacre at Fort Hood would damage his diversity efforts! This leadership relieved the Judge in the trial of the Ft Hood terrorist for enforcing military shaving rules on the terrorist. And they called that obvious terrorist massacre work place violence deliberately depriving those killed and their families of deserved benefits.

Unimaginable in our past, we have leaders who considered medals for not shooting, and a medal for risking carpal tunnel syndrome that outranks the time honored bronze star for valor. This gaggle actually lost graves of our warriors at Arlington cemetery and have attacked the benefits of America's nobility – our veterans. I don't know where the term girlie men came from but it perfectly describes many in the past administration and their military leaders. Obama, Panetta and their military leaders were a perfect storm for the disaster that assaulted our readiness with women in the foxholes and homosexuals in the barracks.

I have said, and many men agree with me, that Adam's rib was the greatest investment in human history. Why? Because God then gave man woman, a different creature, not unequal, different, who is not designed to compete with him but to compliment him. God did not design mothers to kill; and though a very few do, the exception should never be the rule.

You are messing with God's design when you deliberately make killers out of women. Feminist et al, get over it. It is not discrimination to accommodate God's design, it is acknowledging His will – it is wisdom; and it is science.

Despite «Kill Bill» and other Hollywood visuals of females pummeling men, women for the most part are not designed to kill. And they will not be good at it. God designed them to produce life and nurture it, not destroy it. They don't belong in the trenches of the NFL or in the octagon in ultimate fighting; combat is the ultimate ultimate fighting and they don't belong there either.

It is difficult to teach some men to kill but they have no choice. Imagine a draft and a nation forcing our women into close combat killing units. Visualize what will happen to women POWs not to mention homosexuals captured by likely enemies. We have heard of the man who sent his wife downstairs to check on a possible burglar (I actually knew such a man) we are becoming a nation like that man, a wussified nation. There will always be burglars, (international thugs) most of whom are male, and they should be confronted by males. We have heard that we should not mess with mother nature. Worse yet, we should not mess with mother nature's father, God. We have usurped God on marriage, and his design of the sexes. We are in the swirl around the drain and we better get back in line with Him or this will not end well.

INTRODUCTION
Statement of the Case

I went so far as to reflect on one campaign in particular, the Chosin Reservoir, forty years ago. December 1950, North Korea. Probably the greatest—one of the greatest epics of all times. 1st Marine Division, confronting eight Chinese divisions spread out over a long thirty-five to forty-mile linear disposition, north-east—north-south. In extreme cold, −25. Winds out of Siberia bringing the wind chill factor down into God knows what. Mountains. Constant attacking—they attacking us, we attacking them. For days, night and day. Death all about. Frostbite, inadequate clothing. I said, 'Suppose we had 15% women, 20% women.' My supposing led me to say I wouldn't be here. I guess Kim Il-sung would be taking care of my bones along with everybody else's in North Korea.
— Gen. Robert Barrow testifying before the Senate Armed Services Committee, 1991

COMBAT IS AN EXTENSION of man's unique human nature, and because of woman's unique nature she never has succeeded, is not now succeeding, and never shall succeed in this arena. Thousands of years of recorded history and an ocean of anecdotal evidence support this obvious observation.

Yet here we are today seriously discussing whether or not we should institutionalize women in combat in the United States military.

In 1991 former Marine Corps Commandant General Robert Barrow, testifying before the Senate Armed Services Committee, stated passionately that we must not do it.

General Barrow was blunt. "Women can't do it—nor should they be even thought of as doing it."

But the proponents of women in combat declare that it should be institutionalized, and their arguments may be summed up as follows:

1. It is fair to allow women, and unfair not to allow them.
2. Women are succeeding in combat already.
3. Women in combat will increase combat effectiveness.

I contend that a fishing net holds more water than these arguments, and were we having artillery duels across the Mississippi River with a determined enemy that is our equal, these fanciful notions about institutionalizing women in combat would be tossed out the window faster than you could say, "Whose dumb idea was this?" Furthermore, it would be the women themselves who would do the tossing.

Of course Barrow's warning, along with many others since, has gone ignored. In January 2013 then Secretary of Defense Leon Panetta lifted the ban on women serving in some combat positions, allowing women to serve on the front lines for the first time. Secretary of Defense Ashton Carter went one step further in December 2015 by opening all combat units to women.

As a disclaimer let me state now that I am neither scholar nor journalist. I am but a country philosopher. My own short-lived, youthful experience in the military was foolish, embarrassing, and disgraceful, and having never been in combat I claim no moral authority to speak on the subject. But I do dare to speak by the same authority as that of the little boy who cried "The Emperor has no clothes!" My authority is common sense. Furthermore, General Barrow, a highly decorated combat veteran who fought in World War II, Korea, and Vietnam, possessed a superabundance of moral authority. I speak from the ramparts of his shoulders.

Inspired by the late General Barrow, I will undertake in this short treatise to demolish the styrofoam foundation used to support the most horribly misguided idea in the history of the world: the institutionalizing of women in combat.

Make no mistake. In no way did General Barrow denigrate women; neither did he say that women should not have a role to play in our defense apparatus. I will add that women can fight and have always fought since the beginning of the human race.

But we are not debating whether or not women can fight but rather whether or not the United States military should plan ahead of time to include women in combat in America's future conflicts, battles, and wars.

I contend that to institutionalize women in combat is unnatural, foolish, and immoral. I furthermore contend that only because of the convergence of the evolution of *Progressivism* and a window in time that makes the idea seem plausible are we discussing a paradigm that at first, second, and third glance has its roots in basic human nature and is seconded by five thousand years of recorded history.

Futhermore I contend that to have been in uniform does not qualify you to overrule or disregard human nature, and I posit that many combat veterans who defend institutionalizing women in combat have either a very loose definition of what combat is or they are exaggerating their own experiences. I further contend that so called expert proponents of institutionalizing women in combat, be they in or out of uniform, either do not possess a basic understanding of human nature, are historically illiterate, are ideologues, are political climbers, are on the take, or any combination thereof.

This is an issue of first principles and as such anyone possessing common sense and an open mind may understand it. I do not need to be a mathematician to know that $2 + 2 = 4$. I do not need to be a sailor to know that the ocean is wet and that I can drown in it. Neither do I need to be a geologist or oceanographer to know that the Pacific Ocean is not dryland, and if such experts should point out its many islands and bury me in an avalanche of studies as proof that it is fact dryland, I shall dismiss them out of hand and don a life preserver.

But to prove such nonsense you would certainly have to rely upon long and complex studies, which is what the proponents of institutionalizing women in combat have produced, all supported by an ideological media. But all together they amount to nothing more than a noise and light show designed to drown out the voices of common sense and distract a prosperous and secure public that really just doesn't care.

Lastly I have given no anecdotal evidence from within the US military nor of modern combat that might support my point, though they are legion. Qualified men and women have done this and do this daily, only to be ignored. Rather my objective is to point out the obvious to encourage the intellectually honest to reflect upon first principles of human nature and to reach the obvious conclusion themselves. My hope is that those men and women that have served in any capacity in the United States military will reflect upon their own experiences and have the courage to speak out more boldly and force a complacent public to realise that the excessive feminization of our military over the last several decades has profoundly weakened its combat culture.

In Part One of this book we'll begin with human nature and then show that the idea of institutionalizing women in combat did not materialize out of thin air, but rather is a *bad idea* born of other *bad ideas* that have been gradually seeping into our culture for three hundred years. We'll review the nature of combat and then debunk the arguments made by the proponents of institutionalizing women in combat.

In Part Two we will conduct a survey of human nature and observe how since the 1960s our culture has been turned inside out, and how women in combat is but the latest radical societal change on a list that includes the denigration of the homemaker, celebration of promiscuity, mass abortion as contraception, spiraling divorce and illegitimacy rates, a bloated welfare state, the legitimization of homosexuality, and now so-called gay marriage. These are germane evils that emanate from the same source and all strike directly at the fundamental building block of the human race: the *natural-rational-family*.

Finally, based on historical context and a common sense understanding of human nature, Part Three will deal specifically with the reasons why we must not institutionalize women in combat.

In order to establish some context, the following movies are highly recommended viewing for the vast majority of us who want to have a better understanding of what combat might be but have never so much as smelt the gunpowder, much less choked on it.

> Movie: *Saving Private Ryan*, 1998, American
> Director: Steven Spielberg
> Setting: 6 June 1944, Normandy, France, D-Day

> Movie: *Tae Guk Gi*, 2004, Korean
> Director: Kang Je-gyu
> Setting: Korean War, 1950–1953

Movie: *My Way*, 2011, Korean
Director: Kang Je-gyu
Setting: Begins in Japanese-occupied Korea in 1928 and ends on the beaches of Normandy, 1944

Movie: *9ᵗʰ Company*, 2005, Russian
Director: Fedor Bondarchuk
Setting: Final months of Soviet Union's occupation of Afghanistan (1979–1989)

Movie: *Das Boot*, 1981, German
Director: Wolfgang Petersen
Setting: On board a German U-boat in the Atlantic and Mediterranean during World War II
Note: Watch the 1997 director's cut re-release.

Movie: *Hacksaw Ridge*, 2016, American
Director: Mel Gibson
Setting: Battle of Okinawa, 1945

No movie can adequately capture the thrill, horror, or terror of combat, but I imagine these movies have come as close as any.

For a more philosophical and technical understanding of combat, I highly recommend reading the United States Marine Corps' *Warfighting*, its doctrine on waging war.

The common thread throughout this book is the understanding that though times change, and with them conventional thinking, one thing does not change and that is human nature. So let us begin there.

PART ONE:
HUMAN NATURE, HISTORY, AND *BAD IDEAS*

I. The Root of The Issue
Unchanging Human Nature

ALL THAT MANKIND IS and does, and consequently all of human history, is an extension of human nature, and this nature is manifest to anyone who would see it. Mankind is like Earth's woods, forests, and jungles. A close study will show them to be wonderfully complex and wildly diverse, yet from the air they look like the same bunch of trees. Likewise man, and the groups of men that we call *peoples*, appear to be infinitely diverse. But if we take a step back, they are all bound by the same universal and unchanging nature.

I say unchanging, but how could I know for certain? I wasn't present at the *Big Bang*, and had I been I would not have had the patience to watch the slow emergence of man and his evolution into what he is today. Neither was I present when God shaped man from the clay and breathed into him a measure of Himself. However, I am certain that whether we are the creation of God or the result of *Accident*, we have changed so slowly that we are for all practical purposes unchanging, and are now as our ancestors have always been and our descendants will always be.

Yet men are as peculiar as the *peoples* into which they are born, and it is not so difficult to distinguish man's universal and timeless nature from his peculiarities or the customs of men that throughout the ages and around the globe have made individual men and the *peoples* of Earth, like the trees and woods, so unique.

One man sacrifices to Zeus, the other believes in the *Big Bang*, but all men would know from whence they came.

One hunts seal, the other raises goats, but all would feed their families.

One dwells in a tent, another in a cottage, but all would protect their children from the elements.

One *people* eat their suppers at 5pm and another at 10pm, but both enjoy eating together and at the end of their day will sleep, though some in hammocks, some on feather beds, and others on straw.

One *people* are famous for their hot tempers and another for their calm, but all are subject to pride, greed, lust, envy, gluttony, wrath, and sloth. All men are moved by love. The exercise of prudence, justice, temperance, and courage will benefit all men for all time.

Language, beliefs, habits, customs, mores, and traditions morph continuously throughout the ages. From the cauldron of humanity *peoples* emerge, rise, fall, and are submerged again.

But through all of this, human nature has always survived intact and defied any attempt to alter it. Let the river rage and flood, driving its banks to and fro, choking itself on the trees and whole woods it uproots. At the end of the day it cannot break out of its confines. It's still a river flowing downhill.

No, human nature has not changed, but something else has.

II. The Root of the Issue
Feminism's Family Tree, One *Bad Idea*
Begets Another

HOW IN THE WORLD IS IT that we are seriously discussing the idea of institutionalizing women in combat?

What is the genesis of this most foolish of ideas, and how has it come to be embraced by so many in positions of authority or accepted with such nonchalance by the rest of us?

I posit that the root of the *how in the world* can be found several hundred years ago at the birth of *Progressivism*, and I furthermore suggest that if *Progressivism's* natural enemy, *Conservativism* (whatever this term may mean), can ever figure out this *how in the world* it will be akin to finding the goose that laid the golden egg, Excalibur, or Aladdin's Lamp. It will be the kiss that woke Snow White. Suddenly *Conservatives* will remember what their great-great-grandparents forgot and what their grandparents never knew, namely, the *what in the world* we *Conservatives* ought to be conserving.

For generations *Progressivism* has had whatever *Conservatism* is on the ropes, winning bout after bout. When *Conservatism* does get in a few licks and seems to have *Progressivism* on its heels, you can bet your bottom dollar that in no time *Conservatism* will be trapped in a corner covering itself, just trying to avoid the knockout blow and suffer through to the next bell.

When the bell does finally ring, *Conservatism* will raise its gloved fists and cry out to God once more: "*How in the world* is it that *Progressivism*, with all its kooky ideas, keeps whipping me?"

The *how in the world* is to be found in the brain orgy wherein *Progressivism* was conceived.

♦ ♦ ♦

Several thousand years ago there occurred the first great agricultural revolution, one that would see man for the first time transition from hunting and gathering to settled agriculture. As a consequence, in time,

5

and particularly with the development of irrigation, what we call civilization would emerge along the banks of some of the world's great rivers.

By civilization we mean that man's ideas and society became more complex. Settled agriculture was far more productive than hunting and gathering, and the subsequent population growth combined with permanent settlements produced a surplus of manpower and material that allowed for the leisure to speculate and the resources to put that speculation into practice. Now amidst all this farmland teeming with yokels there would be towns and cities full of all manner of special groups and specialists. One group would rule, another would fight, and another would speak the will of the gods. One man would forge swords, another tan hides, another fashion jewelry. Others would pack these goods into their boats or onto the backs of beasts and carry them to towns nearby or far off to trade, and some men would keep track of all the new things that were being done and made.

But in all this commotion man's nature did not change from that of his hunter-gatherer forebears. The desire for security remained the universal constant, while the lust for power, glory, and women drove certain men as they always had, who would in turn drive events. Man still made love, and he certainly still made relentless war. He just did so from a chariot in a sea of shaking bronze-tipped spears, and in the name of a great king who would take the leash off his victorious soldiers to plunder his enemies. And oh what plunder was to be had in this bountiful new age!

As for the king, plunder was all fine and good and necessary to pay the troops, but the real prize was becoming master of the new system. There was tribute to be exacted, trade routes to be controlled, taxes to be levied, and rents from farmland to be collected.

Wherever man went, whether to trade, make war, or settle down he carried these new ideas with him, and thus civilization spread. Millennium by millennium, century by century the hunter-gatherer was slowly squeezed until he hardly had a scrap of land left to graze his ponies.

Just as all hunter-gatherers in all places and times would have recognized each other, so would this new settled, civilized man. From the emergence of the first civilizations in Mesopotamia and Egypt up until the modern age itself, the face of civilization really did not change that much.

There were woods, fields, orchards, or pastures where the bulk of men toiled, and towns, cities, plantations, manors, castles, etc., where the specialists lived and worked. News traveled at the speed of horse or wind, and muscle, wind, and gravity moved that which was moved. The gods explained the inexplicable, and the priests explained the gods.

But then finally came the 17th century AD and there occurred another paradigm-shattering event, an historic event of such magnitude that it would radically alter the way civilized man had lived and thought since long before Narmer sat upon the throne of a united Egypt.

I speak, of course, of the *Scientific Revolution.*

This event was led by men whose names still ring bells of recognition in practically everyone: Copernicus, Descartes, Kepler, Galileo, Bacon, Newton, and others. Truly deserving of the title, the *Scientific Revolution* began the demystification of nature and the radical acceleration of man's understanding of the physical world that continues to this day, and that has led to a transformation in the life of man every bit as radical as the one seen on the banks of the Tigris and Euphrates some fifty-five hundred years ago.

The Birth of The Industrial Revolution

The *Scientific Revolution's* radical transformative power was unleashed in the 18th century by its child, the *Industrial Revolution.* Begun in Great Britain, this revolution has since swept the world.

With a vastly improved understanding of the physical world came the ability to manipulate it to an extent unimaginable to our ancestors. The ancient agrarian order established on the banks of the Tigris, Euphrates, Nile, Indus, and Yellow Rivers so many thousands of years ago, and still the order of the day in West Tennessee in 1900 AD, would be upended by the *Industrial Revolution* and replaced by a radically new economic order that would manifest itself in the cities and suburbs that so many of us today call home.

This new economic order has had a dramatic impact upon virtually every aspect of human life in the last three hundred years, but its effect upon the human home and family has indeed been unprecedented.

But let's put the human home on hold for the moment because while the *Industrial Revolution* was transforming the landscape and how *peoples* lived, its twin, the *Enlightenment,* was effecting a change in the human mind that would impact human life even more profoundly, and devastate entire nations and homes beyond reckoning. Together, the *Industrial Revolution* and the *Enlightenment* have led us to the strange debate at hand.

♦ ♦ ♦

By contrast to the *Industrial Revolution,* that revolution we term the *Enlightenment* was essentially a theological—I mean, ideological transformation.

Early modern science made it clear that both the world around us and the human body itself could be understood, and as the world was made more productive by this understanding, human health was likewise improved. For example, as it turned out, the heart was a pump circulating blood throughout the body. Like the invention of the wheel it was only obvious after the fact.

Hence, the great theologians—I mean, philosophers of the *Enlightenment* asked, "If the natural world and our bodies could be understood, controlled and even improved, then why couldn't man himself be?" That is, the inner man, the thinking man. And if he could be improved, why not human society?

And thus was born a new religion, one that has ever since waged relentless war on older worldviews like Christianity, Islam, Hinduism, and Buddhism that have presumed to ask and answer man's heady questions. In this treatise I am going to call this new religion *Progressivism* and its adherents *Progressives.*

I call *Progressivism* a religion because like all religions its doctrine is built upon some *leap of faith,* some unprovable assumption which supports and leads to everything that follows. For the Christian that leap of faith is the belief that Jesus Christ is the Son of God. For the Muslim it is the Shahada, *There is no God but Allah, and Muhammad is his prophet.* For the *Progressive* it is an unquestioned belief in a new pantheon of gods spawned by the *Enlightenment,* which I'll call *The Great*

8

Assumptions: Reason, Individuality, Freedom, Equality, Perfectibility, and others.

The gods *Freedom* and *Equality* in particular have given *Progressives* two commandments that they have embraced with childlike faith:

All men are created equal.
All men are born free.

With the aid of the *Industrial Revolution's* transformation of the economy, these commandments have given birth to absurdities (such as the debate over institutionalizing women in combat) because both of these statements are manifestly untrue, can only be true when qualified, and are themselves irreconcilable. Where there is freedom there is no equality, and where there is equality there is no freedom.

Men may be created equal in the sight of God, but they are otherwise as unequal as trees. Clearly, no baby is born free, but is utterly dependent upon its mother and limited to some extent by the society into which it is born. But in practice *Progressives* have reinterpreted these statements as injunctions, i.e., *man ought to be equal* and *man ought to be free.*

The Great Assumptions conceived, in their unfettered thought-orgies, numerous baby-isms including Communism, Socialism, Capitalism, Libertarianism, Nationalism, and Nazism, and have resurrected a few old ones such as Republicanism and Democracy, each new little godlet convinced that it was the child of destiny that would lead humanity to the promised land—wherever that is.

But when *Freedom* and *Equality* hooked up they spawned *Feminism,* the single nastiest baby-ism of all time, and from a list that includes Communism and Nazism, that is a bold statement. After all, the Nazis imagined only that there was a pure German or Aryan race and that this race *ought* to rule the world, while less exalted races would be obliged to make way for more Germans. The Communists merely tried to reduce men to the material level and then make them all equally poor.

Both ideas would lead directly to the deaths of tens of millions and a gale of human suffering, but both are horrible misunderstandings when compared to the damage that *Feminism* has wrought upon mankind. The Nazis and communists may have rip off large healthy branches like a trackhoe cleaning up a treeline but *Feminism* has poisoned the roots. *Feminism* split the atom. It broke the fundamental building block of humankind and turned it against itself: It turned woman against man.

Based on the unquestioned belief in the sovereign goods of *Freedom* and *Equality*, and fueled by the natural self-centeredness of both sexes, *Feminism* has declared that woman *ought* to be independent from and equal to man.

Thus *Feminism* wants women in combat. If women are excluded from combat then both tenets of its faith are violated. Woman is then not only not equal to man, she is dependent upon him and thus not free. For *Feminism* this is nothing less than slavery.

But the idea of women in combat is bad science fiction. How, then, has *Feminism* been so successful in promoting the idea? If the idea flies in the face of human nature and common sense, it must be in part because of the opposition's weakness—that opposition naturally being *Conservatism*. And what is the weakness within *Conservatism* that has made it a pushover in the face of such nonsense? Nothing less than *The Great Assumptions* themselves.

Conservatism continues to lose to *Progressivism* on this issue and others because *Conservatism* has embraced the intellectual foundations of its putative rival. What we call *Conservatism* today loses because it, too, worships *The Great Assumptions*, for it is itself but another child of the *Enlightenment*, another baby-*ism*.

If whatever *Conservatism* is, is to turn the tide of battle, it must reform itself first.

Conservatives must understand the context of the debate, reject *The Great Assumptions* as sovereign, argue unapologetically from the vantage point of a rational understanding of human nature, and finally agree upon what is indeed sovereign truth. Only when we finally remember what we *ought* to be conserving will we be able to turn the tide against the absurdity of institutionalizing women in combat, and against *Progressivism* generally.

III. Why The Bad Idea of Women in Combat Seems Plausible
A Window in Time

WHAT IS THIS COMBAT THAT General Barrow and his camp, and the proponents of institutionalizing women in combat are debating? First, I am going to assume that we all agree that women had no meaningful role to play in combat prior to the use of gunpowder in battle—that is, back when combat meant fighting with a club, spear, sword, or even with bare hands. Yes, spearmen, bowmen and slingers fought from a distance, but a charging enemy could close in on them in seconds, forcing close-quarters combat, and the firing of such missiles with lethal effect in the maelstrom of battle required more than a little strength and endurance. No, women could not have fought effectively at Marathon, Cannae, or Agincourt, the rare exceptions being just that.

But with gunpowder came weapons that could kill the biggest man from a distance, that could render body armor almost useless, and that required relatively little strength to fire. Setting aside centuries of variables such as the difficulty of operating an arquebus versus an M-16, there still remained the problem of close-quarters combat in which trained women have virtually no chance of success against trained men.

If anyone has an inkling of doubt whether women can perform well in close combat against men I strongly recommend that person spend a couple of hours trawling YouTube for UFC fights.

It wasn't until the invention of the machine gun and its widespread use in World War I (1914–1918) that close-quarters combat became less the order of the day. Setting aside the question of whether women could have survived and performed well in the trench warfare that was the hallmark of that war, there still remained the problem of the forced march.

Prior to the mechanization of warfare (trains, trucks, airplanes, and helicopters), the infantryman traveled across land and continent on foot carrying his gear. Let's assume that since the advent of gunpowder the infantryman has carried gear that weighs 50 pounds in excess of his own body weight—probably a very low estimate. We'll also assume that these

11

infantrymen were required to march 15 miles per day, day in and day out. This too could be low. There may be a few fit women in their twenties that can walk the length of Tennessee at the rate of 15 miles per day while carrying 50 pounds and still be fit to fight in the afternoon, but not many. But when you add the problem of the inevitable forced march where much greater distances must be traversed, at speed, and in harsh and desperate conditions, there are, for all practical purposes, none that can endure it. History certainly provides few examples of women that did.

To be clear, I am not talking about extreme hiking or military training. I am rather referring to forced marches by foot, in combat theaters, under extreme conditions, while maintaining combat effectiveness. Examples of such marches would include Hannibal's Crossing of the Alps, 218 BC; Alexander Suvorov's Swiss Campaign, 1799–1800; and the fighting retreat of the U.S. 1st Marine Division during the Battle of Chosin Reservoir, 1950.

I suppose there are a few exceptionally tough women who could have kept to their feet and *under arms* in these terrible campaigns, while men in their prime were dropping like flies—but not enough to matter.

Surely no serious person believes that women in combat could have been institutionalized when hand-to-hand combat with men was the norm, and the forced march common, two realities of combat that find their roots in prehistory. Honestly, anyone that will tell me that women could have had a meaningful role to play in the fighting beneath the walls of Troy would as likely argue that the ocean isn't wet, and I just don't know how to talk to such people.

So what has changed about war that has given *Progressives* the hubris to propose and enact such an absurdity as institutionalizing women in combat? Is combat no longer fought at close quarters with wild-eyed, adrenaline-soaked men in their twenties? Does it no longer require feats of endurance that tax these young men in their prime to the point where lying down in the snow to freeze to death would be thought sweet release?

This must be the case. The ancient paradigm of combat must have changed thus making it possible for women to perform well in it. Or weak or old men for that matter.

But if there is one thing history has taught us, it is that history never ends. It cannot end because it is driven by man's nature, which remains unchanged and unchangeable, and thus will produce again and again what it has in the past—in this case, war. As a consequence, man's actions can be predicted, the degree of accuracy depending on how abstract or particular the prediction.

Will the United States fight another war? Almost certainly, yes.

Will the United States fight a war with China? It is vastly more likely than our fighting a war with Norway, though by no means certain.

If the United States fights a war with China, what day will it begin? Your guess is as good as mine.

But whether the issue is abstract, strategic, or tactical our generals, admirals and those charged with the defense of the realm must at all times peer into their crystal ball and determine what tomorrow holds. That crystal ball is the past.

We study the past in order to anticipate the future that we may act wisely in the present. Concerning war, we have thousands of years of history to help us anticipate the future, and the history of war has consistently been one of contingency and brutality, fraught with consequences sometimes horrific and occasionally apocalyptic. Common sense tells us that combat is an extension of man's unique nature, and history tells us only men have ever been able to succeed on the battlefield. Isn't it wise to assume that it will always be thus?

But if there has indeed been a paradigm change that now allows women to succeed in battle against men, when did it begin and what evidence proves it is permanent? Otherwise, I contend that *Progressivism* has only been able to canonize this important doctrine because an extraordinary set of circumstances has allowed it to be passed off as plausible.

We are seriously discussing women in combat today because of the short-term convergence of unique and, in some cases, unprecedented historic, cultural, geographic, and economic factors that are allowing *Progressivism* and *Feminism* to create a believable illusion. Consider the following:

- These United States are flanked by the world's two largest oceans, are warmed by the thick quilt that is Canada, and sit firmly on the governmental and economic basket case that is Mexico.

- Somewhere down there is South America, which appears to be about as oblivious to us as we are to it. Geographically speaking we're about as secure as any country on the planet.

- We have not seen war on a large scale on our own soil since 1865 and haven't seen mass casualties since 1945, which could account for the strange disconnect between Americans and the nature of war. Even Southerners have forgotten Sherman's March to the Sea.

- The Cultural Revolution of the 1960's was real, and for generations now it has profoundly altered our collective sense of what is right and wrong, natural and unnatural.

- The crushing of Nazi Germany and Imperial Japan, the bleeding to death of the European colonial powers, and the near breaking of the Soviet Union left the United States of America the last man standing after World War II, muscled and virtually unscathed. Our economic model became that of the free world, and for decades afterwards we enjoyed an unprecedented economic hegemony over the entire globe that has only receded in recent years.

- With the destruction of the old imperial paradigm in World Wars I & II, the likelihood of war between peer Western powers was reduced to, and remains, almost nil.

- In 1945, only the Soviet Union remained in a position to go toe to toe with the United States and its allies, but the bankruptcy of its communist faith had by 1989 destroyed it from within, leaving the U.S. the first sole global superpower in history.

- The *Scientific Revolution* of the 17th century has continued to spawn new technological revolutions up to this day that, combined with the great wealth it has also produced, has given the Western powers and the United States in particular a military supremacy over all other likely enemies that is as complete as any general could have dreamed of a hundred years ago.

- We sweep third-world armies aside, our most powerful potential competitors dare not challenge us on land, at sea, or in the air, and there is no corner on Earth where the rogue can hide from our military or diplomatic reach.

- When we do fight, our casualties and material losses are strategically insignificant. Technology continues more and more to shield us from the enemy and expose the enemy to us. Flight has radically reduced the fatigue of mobilization and the march, and our doughboys in the trenches of World War I would have envied the comforts their frontline counterparts enjoy today. American soldiers rarely succumb to thirst, starvation, exposure, or untreated wounds and are seldom taken captive.

Thus a window in time has opened that has allowed *Progressivism* the luxury of putting women in combat (or at least in combat units) without great risk to the security of the nation and to thus promote the notion that woman is the equal of man in this realm when common sense cries to heaven she is not.

If we recognize the fall in 1989 of the Soviet Union, our last serious challenger, as the opening of this window, then this new paradigm has existed for less than thirty years—about how long it takes history to yawn.

During this time the United States has fought major wars in Iraq and Afghanistan that the proponents of women in combat use as evidence that women are indeed suited for combat. But are these wars acceptable case studies? Let's put this in perspective.

In our wars in Iraq and Afghanistan between 2001 and the end of 2015, we lost a total of about 7,000 dead (and counting), approximately less than two servicemen per day.

In Vietnam we lost a total of 58,209 between 1955 and 1975 (26 per day). We lost 16,592 in 1968 alone (45 per day).

In Korea we lost 36,516 during a much shorter period of time, 1950 to 1953 (45 per day).

In World War II we lost 405,399 between 1941 and 1945 (416 per day).

In World War I we lost 116,516 between 1917 and 1918 (279 per day).

In the War Between the States we lost 625,000 between 1861 and 1865 (599 per day), and this was out of a population of only 30 million.

Now a cop on a beat grappling with a knife-wielding crackhead is living his own little hell, and the Marines who fought in Fallujah in 2004 have more than an idea of what German and Russian soldiers went through in the urban hellscape of Stalingrad, quite possibly the most brutal battle fought in modern history. They must not imagine that I am attempting to diminish their personal experiences. I am simply comparing apples to apples. Fallujah, hellish as it was, does not compare to Stalingrad, where between August 1942 and February 1943 roughly 800,000 German and Russian soldiers were killed at the rate of 3,800 per day. Neither can Iraq and Afghanistan, objectively, compare in scale to previous wars that we have fought or a staggering number of other wars for which we have historical records. That some women have performed admirably in firefights in Iraq and Afghanistan is no argument in favor of disregarding common sense, ignoring history and institutionalizing women in combat.

This window in time is too short, our wars in Iraq and Afghanistan too lopsided, and our military advantage at present too extreme to be used as proof that an idea as thoroughly unnatural as women in combat is wise or harmless, or that our military supremacy is permanent. On this point history is clear: Relative economic power ebbs and flows and sometimes crashes, and technology is highly diffuse. Our military supremacy is not permanent and one day victory will once again require iron resolve, muscle, endurance, ferocity, and a river of blood. Those that have dared to fight us in recent years would certainly testify that it still does, and thousands of years of history testify that it will be so once more even for us.

◆ ◆ ◆

But, as proponents might argue, women have been in combat within this window of time and before its opening, including Stalingrad. There are many other examples. Circumstances aside, isn't that some proof that in spite of the historical record the US military should institutionalize women in combat today?

IV. EXCEPTIONAL TIMES
DO NOT NEGATE THE RULE
Women in Combat is Unnatural

LET'S SAY THAT WE COULD go back in time to 6 June 1943 and recruit 500 of the stoutest, most hot-tempered, high-spirited American women that then existed and for a full year put these women through the most rigorous training course that tough women can possibly endure, turning them into the toughest soldiers women can be. Now fast-forward to 6 June 1944 where our 500 women soldiers have been scattered throughout the Allied invasion force that is about to descend upon Normandy in what will be known as D-Day. A month later any number of these women would still have been alive and could say that they had indeed been in and survived combat. Not the worst ever, but right up there.

Now when former Marine Corp Commandant General Robert Barrow said, referring to combat, that "Women can't do it," he did not mean that just because a woman is in combat she folds like a rag doll or vaporizes. He was stating what he believed to be an irrefutable truism which (because it is a truism) will have exceptions.

It is impossible to have meaningful debate about human existence without resorting to truisms because for any general characteristic of human life there appears to be an infinite number of possible variants and variables, and if we cannot accept general truths because of these exceptions then there can be no practical wisdom.

That the pride of one proud man does not appear to lead to his destruction does not make untrue the proverb, *Pride goeth before destruction, and an haughty spirit before a fall.*

The truth is that we live our lives guided by truisms, rules of thumb, and probability. To live by the exceptions is nothing less than gambling, and the gambler always loses in the end—as a rule.

Why is it then that for so long *Progressives* have controlled the debate by arguing the exceptions? *Conservatives* come to the table armed with what we think is common sense and a clear set of rules—believing that right is right and wrong is wrong, while the *Progressive* comes armed with exceptions to our rules and somehow wins the day.

The *Conservative* says abortion is wrong. The *Progressive* asks, "What if modern medicine declares the life of the mother to be in danger" or "What if the baby has Down Syndrome? What about pregnancy by rape or incest?"

If the *Progressive* doesn't thereby convince the great mass in the middle that abortion is OK because of the apparent exceptions to the rule, he so shakes their confidence that they'll take no side at all.

The *Conservative* says that combat is a man's world and is no place for women. The *Progressive* replies, "But what of the women that have been in combat or are in combat today? Obviously they can endure it." Checkmate!

And the mass in the middle says, "Well, I'll be. I guess they can."

No—abortion is wrong. That some people can think of limited circumstances in which abortion might not be wrong does not make it right and is no justification for the current situation in which abortion is used almost exclusively for the purpose of birth control outside of wedlock, which was the objective of *Feminism* all along—think *Sexual Liberation*.

No, institutionalizing women in combat is wrong—and dumb. That there are examples in history of women that have survived and performed well in combat is no justification for ignoring nature and doing something that in the United States today is wholly unnecessary.

♦ ♦ ♦

In Russia World War II is known as the Great Patriotic War and it is a prime example of when it might have been smart to institutionalize women in combat or to put them in units that get awfully close to it. Between June 1941 and May 1945 the Soviet Union lost a staggering 26,000,000 dead (the US lost 405,399 in WWII) and this number may be low. Setting aside the Soviet Union's communist ideology, women were thrown in the breach at the Battle of Stalingrad and other battles because Soviet men were being slaughtered by the tens of thousands by a Nazi Germany that, if victorious, might literally have systematically exterminated millions more (women and children included) in order to create more Lebensraum for more Germans.

But the United States is not facing such a threat today or even the nightmare that the South endured between 1861 and 1876. Neither are we like the Israelis, surrounded by, and within artillery range of *peoples* that consider their state's existence an existential threat.

Yes, a woman will defend her children to the death and even fling herself upon the man that is beating her husband. She'll load muskets and hand them up to her man on the wall of the fort that is being assaulted by Indians who mean to scalp them all. Yes, governments suffering extremely high rates of casualties in their armies may call upon women in order to free up men for the front. Some *peoples* may be so surrounded and oppressed by enemies that their governments feel compelled to keep women soldiers in reserve and give them some combat training. But governments do this when they are desperate, not because they believe women are effective in combat against men or because they are indifferent to the mangled bodies of wives, daughters, and sisters.

All exceptions to the rule combined, along with every example of women in uniform (in combat or not), add up to a drop of water in the ocean that is the history of combat. Multiplied a hundred times they make no compelling case for institutionalizing women in combat in the American military today.

We should live by the rule, not by the exception.

V. FREEDOM AND EQUALITY
False Justification for Women in Combat

SHOULD A FEMALE HONEY bee ever possess a minute of conscious thought, and should it occur to her in that minute that the females of her kind do all the work, would she have cause to complain? Likewise, should a male honey bee lament that it is not fair that he should always be ruled by a queen, or that he himself cannot become king and, at most, can only hope to become the queen's consort and then die in the act?

Should a squirrel envy a turtle's hard protective shell or a turtle the squirrel's great speed? And if they did, what good could come from such envy? They are, after all, what they are because it is the inescapable nature of things.

One of the great inconsistencies of *Progressivism* is its simultaneous faith in and rejection of the natural world. On the one hand it teaches that everything is a product of *Accident* or the *Big Bang* or *Chaos* or whatever you want to call the unknowable *Beginning* that could not know itself. *Progressives* are well known for their defense of the natural world and particularly of animals (especially if they are cute). *Progressive* fundamentalists will go so far as to tell us that to eat meat or wear skins is murder, and, if they can, they will force industry to go to great expense to protect the habitat of any given species that it may live according to the needs of its nature.

Yes, man is the most intelligent of the animals, but like the bee, squirrel, and turtle, he is nonetheless a product of nature.

OK, fine. But what of man's nature? Does he not have an animal nature like that of the bees, squirrels, and turtles? Will not observation reveal this nature as it does the natures of other animals? And once identified, shouldn't we accept man's nature for what it is? We do not, after all, lament that the cow must nurse the calf.

It would be manifest to alien observers from another planet that the male and female of our kind have unique natures, just as the female and male bees do. Our alien observers would agree that our kind is propagated via intercourse, the human embryo being conceived within the female and growing within her for about nine months until birth, after which time it is sustained by means of nursing at the same female's breast until weaned. But would it not be equally manifest to them that the human female is also the primary caregiver to her young for years afterwards? Would it not be equally manifest that the male is the builder of our kind? For virtually everything that stands has been made by his hands. More to the point, would our alien observers have any doubt that the male makes war on his fellow man?

Why then does *Feminism* lament that women cannot make war alongside man and demand that she be allowed to? Why can't it accept the manifest reality of human nature, a product of the same *Accident* that produced bee-nature, squirrel-nature, and turtle-nature?

Well, because it's *just not fair* that they are excluded, because woman *ought* to be equal with man and *ought to be free* to do what she wants. Thus saith the gods *Freedom* and *Equality*. Amen.

Fine and dandy. But blast it here comes reality smashing up their tidy new paradigm. For what is this thing called "combat" that must submit to the *Feminist* fairness doctrine?

War, and combat specifically, is state-sanctioned brutality and deceit, the stakes of which are life or death, freedom or slavery, sovereignty or submission. It is also manifestly an extension of man's unique human nature. How can fairness possibly be applied?

Imagine a *Captain* of men upon the battlefield. His enemy counterpart has tricked him into making a false move and he now finds his company exposed and pummeled without mercy. He must retreat, but must gain time to do so. He looks about him and meets the eyes of two of his sergeants. One is insanely brave and somewhat fatalistic. The other is a good soldier and a fine administrator who looks forward to taking over his father's muffler shop when he retires from the army. In his eyes the *Captain* sees fear beginning to set in.

The *Captain* now orders the first sergeant to stay put and fight to the death, but to be certain not to die until the company has withdrawn.

What is fair about that? Shouldn't the *Captain* have flipped a coin? Shouldn't he have taken a vote? Or couldn't he have called out to his adversary across no-man's land and said, "Hey feller! That was a low-down dirty trick you played! You can make up for it by letting us go and we'll duke it out tomorrow. That's only fair!"

Fairness and mercy are all fine and good in war when victory is assured or they can be afforded. But make no mistake, if our *Captain* is required to "be fair" on the battlefield, he will face an opponent who is not, and who will exploit every weakness in our *Captain* in order to destroy him.

Militaries have always been and must always remain dictatorships. Fine notions of fairness have no place in them and the *Captain* must never be asked to take fairness into account any more than pedigree. On the castle wall he must be allowed to make decisions solely on the basis of what will assure his sovereign victory or at least an acceptable, negotiated peace.

Young *Captains* choosing sides for a basketball game can pick the frail boy with a limp for reasons of fairness, compassion, or playground politics because it is in the end just a game, but the defense of the realm is not.

♦ ♦ ♦

The history of humanity is an endless chronicle of suffering because we cannot escape our human nature or nature itself. Life is hard and painful enough when we are in good health and in possession of a sound understanding of ourselves and circumstances, but we are not always healthy, and at times our understanding and judgment are flawed. At other times we are caught up in events beyond our control. In good times or bad, some degree of suffering is simply the nature of things.

But to willfully disregard the fundamentals of our human nature in good times is to invite bad times. I may imagine that I can fly, strap on wings, and prance about as though I am a bird, but if I jump from a cliff, I will likely fall to my death. That I might survive the fall the first time does not make me a bird or any less a fool. I am a man bound by human nature and this nature is not hard to understand.

PART TWO:
FEMINISM VERSUS HUMAN NATURE, THE FAMILY, AND THE UNITED STATES MILITARY

VI. THE NATURAL-RATIONAL-FAMILY OUR HEALTHY-NATURE

IN THIS TREATISE I am declaring that *institutionalizing women in combat* is the most absurd idea in the history of mankind. My arguments rest upon 1) a belief that human nature is unchanging; 2) a common sense understanding of this human nature; and 3) the testimony of human history, which is itself a result of this unchanging human nature.

Beginning with the debate at hand, let's descend into this human nature, get to the bottom of it, and see if it is not obvious.

We know what *women* means in the statement *women in combat,* but what about *combat?* What it does not mean is man-to-man combat, like that between two medieval knights or UFC fighters. We could be referring to melees between private parties like the Crips, Los Zetas or Halliburton. We certainly are referring to a fight between groups, each sanctioned by an entity that claims some degree of sovereignty. In the context of the debate that we are having in the United States, by *women in combat* we mean very specifically by forethought and planning putting women in either US military ground units that engage in close-quarters fighting or in aircraft and warships that engage the enemy directly. Women in combat should also refer to the assignment of women to military units operating in a given combat theater that would likely be the target of an attack by an enemy combat unit.

But who are these groups that are fighting each other? Throughout history down to the present the human fighting group has been a host of fighting men led by a *Captain.* They are a group because they are drawn from the same *people.* They are fighting in order to protect or advance their common interest against another group. These common interests are possessions, whether physical (survival, family, cattle, lands, trade routes, women, etc.) or notional (ideas, sovereignty, self respect, the respect of others, glory, etc.).

But why are these groups fighting over these physical or notional possessions? They fight because to struggle is a human imperative, competition is a basic human pleasure, and fighting is its most intense and conclusive expression. Struggling is simply what man and by extension his group does. He struggles (fights) for joy, power, glory, and women. He fights to gain that which he needs or wants, or to defend what he possesses. He strives to rule, to not be ruled, or to survive.

He strives to survive because that is the first human imperative. Not all men want to fight but some men always do and must be contended with. Not all men strive to rule, but all *peoples* produce men (or women) that do and they inevitably drive events.

He *must* strive to survive because of other men striving against him, and because the natural world in which he lives and cannot escape is at all times trying to kill him by means of starvation, exposure, injury, disease, and ultimately decay.

Each fighting group is itself a subgroup of a much larger group that we'll call a *people* and exists only because of the *people* from which it is drawn.

A *people* is a collection of human beings led by a *Captain* (be he a chief, warlord, governor, king, etc.) that are identifiable as a *people* because of similarities and commonalities born of shared experiences within a common space that binds the members of the *people* by links of sentiment and shared interest.

The individual members of the *people* share these similarities because of man's natural inclination to imitate and conform to those with whom he shares space and because via intermarriage men and women, in their offspring, will produce a blend of their own kind. This conformity will be manifest in shared language, ideas, and customs (i.e. culture) and given time, will result in a shared ancestry that will be manifest in their physical attributes (i.e. race).

Peoples have always existed and will always exist because man lives in space and time, is born into a *people*, is utterly dependent upon a *people* to survive his childhood, desires the company of his *people*, and will only find security against nature and the aggression of other men and *peoples* as a member in good standing of a *people*.

A *people's* ability to survive within nature that is at all times attempting to destroy its members is dependent upon the bonds that bind them together, namely the bonds of shared interests and sentiment. If these bonds are weak, the *people* risk disintegration or absorption by other *peoples.* If the bonds are strong, the *people* is more likely to survive intact by the lending of mutual aid between its members.

The bond of sentiment is born of man's nature and is as real and consequential in human affairs as is the inclination of wolves to live in packs. It is not a social construction, or a custom subject to change. It always exists or is in the process of coming into existence.

The binding strength of sentiment varies along a broad spectrum. At its weakest end is the sentiment between men who share the broadest cultural background (linguistic, cultural, ancestral). This is the bond that causes a man from Arkansas to sympathize less with Chinese families suffering flooding on the Yellow River than with Austrians affected by flooding on the Danube.

At the other end of the spectrum is a bond of intense, sacrificial love, i.e., the kidney-bond—a bond so intense that one would give a kidney to save the other. Akin to this is the shock-and-wail-bond—that is, a bond so close that the unexpected death of one will bring paralysing agony to the other. The most common of these bonds are those between husband and wife, between them and their biological offspring, and between siblings. However, this bond is common in many other human relationships, e.g., extended family, adoptive family, boon companions, soul mates, etc.

Between these two extremes of broad sentiment and intensely emotional, unqualified devotion are countless degrees of sentiment and a constellation of human bonds, both individual and corporate. For example, there is the bond between two strangers sporting the same Cardinals baseball cap, or the bond between student and teacher. There is the bond a man feels for his political party, or for the insurance company that pays on his claim. There is the bond between the faithful employee and the just and generous employer, or the bond that connects people who have embraced the same religion or idea.

Any one individual may be linked to any number of other individuals or groups, but all bonds are subordinate to "kidney-bonds," the strength of each bond being directly related to the support it lends to said kidney-bonds. The bond that a woman may feel for the other members of her bowling team may be strong and have consequences for her daily life, but it pales in comparison to the bond she feels for the men upon her city's walls who are struggling to throw off the scaling ladders of a fierce army that will slay her husband, sack and plunder her city, rape her, and enslave her children if they can.

Let's suppose her husband, a bowling shoe cobbler, is upon this city wall during this attack. He'll look down the ranks of his countrymen and recognize other men with whom he shares bonds of sentiment. There is the tanner from whom he acquires leather, and the butcher that supplies the tanner with hides. There are his friends from the corner tavern, and the barkeep himself. There is the husband of his wife's oldest and dearest friend. Below them all behind the gate stands a company of regular soldiers, among whom he counts a cousin. But the vast majority of faces he does not recognize. Like him they were all sucked in from the countryside or vomited out of the bowels of the city in answer to the bells that pealed, "We are under attack! Into the city and onto the walls!"

Though this mass of men upon the wall by and large do not know each other, they are all bound by their respective links to the city they would defend. They are all threads in its history and identify themselves with it. They would yet be ruled by their own king and their own laws. They would keep for themselves the labors of their forefathers and in turn bequeath them to their own sons and daughters. These common memories and shared interests will now bind these strangers into a chain that might be strong enough to hold back the tide that now surrounds them, but these commonalities themselves ultimately serve only to protect that which is most important to every man on the wall, i.e., their respective families that huddle within the city.

The protection of their families, each a cluster of intensely tight bonds, will be the ultimate motive for which the men on the wall will stand, fight as one, and hopefully prevail in their life-and-death struggle. If they prevail, this new memory will further bind them together for the remainder of their lives and will serve to unite their offspring for generations to come.

Contrary to *Progressive* doctrine, the individual is not the building block of whatever it is that *Progressives* would see built. Rather it is this family, that cluster of intensely powerful bonds which is the basic building block of the naturally and inevitably occurring *people*, whether a tribe, city state, or nation.

The *people* exists by nature because the human family exists by nature, which itself exists because man and woman naturally possess an imperative to unite, have children together, and defend themselves and their offspring. The natural world and the nature of other men and *peoples* require that a family comprised of a man and woman be united with other families for mutual aid, and by this union they form a *people*. A *people* exists only because of its constituent families, and its purpose is to defend those families.

The family is formed by the joining of two incomplete parts, each possessing a unique nature that is completed only in that union. These two parts are a man and a woman, a husband and his wife. They are drawn together by nature, and this union between man and woman is marriage, which, once formed, produces the *natural-rational-family*, the building block of the *people* of which it is a part. This human family is the fount of security, comfort, satisfaction, human survival and progress.

Contrary to *Feminist* doctrine, man and woman are not equal, nor should they be made equal. When their distinct natures, or rather their *healthy-natures*, are denied, or when they do not fulfill their own, then they produce either a weak family or none at all. When each embraces his or her own *healthy-nature* then the family they produce is strong.

The seat of the family is the naturally occurring home.

It has been said that *home is where the heart is*. This is a fine sentiment, but a more accurate expression would be *home is where the woman is*. Indeed, whether a cave, tent, thatch hut, boxcar, hotel room, castle, or three-bedroom ranch, a man's home is where his woman resides.

In this home the humblest man and wife are sovereign. In this home the ditch-digger rules as king, the scullery maid reigns as queen.

A man by nature longs for this home. Woman by nature longs to make it, and she alone can make it.

A woman by nature longs to be loved, served, and protected by a man in our dangerous and stingy world, and a man by nature longs to be served, needed and adored by a woman.

Both long for children and will sacrifice all of themselves for them.

A man by his nature would lead his woman, and she by hers would follow him.

The man orbits his home to hunt and sow that he may provide her what she needs, and to fight off werewolves, marauders, and zombies that she remain unmolested. Tired and bleeding he returns to the home that she has made for him to be bandaged and enjoy the comforts that she would give. Now rested, he returns to his orbit once again.

He by his nature must look outward to the wider world. That is where he hunts, farms, builds, and fights. If the men of his *people* will come to his aid when he is in need or danger, then he must go to theirs. When the hard thing to be done requires many hands, he must lend his. He is at all times subject to a *Captain* and must answer his call when the territory of his *people* is invaded by another or when his *people* are to invade theirs.

She on the other hand lives in feudal independence within her home, subject only to her own husband over whom she exercises extraordinary power. She who has made the home must at all times look inward to tend to those things that must be done within it and that only she can or will do. She must tend to the young, sick, injured, and elderly of her family. She must garden, gather, store, and prepare meals. She must clothe her family and will manufacture what she can for her own use as well as for trade. She must guard the home when her husband is away and must supervise the labors of her dependents. She is by nature a worker, multi-tasker, and manager.

He is the builder, she is the adorner.

He is the teacher, she is the nurturer.

He is the fixer, she is the comforter.

He teaches us to love our *people*, she to love ourselves. Together they make us love our family.

Hers are the hands that rock the cradle. She makes the first and deepest impressions on her young. It is by her hands first that they are trained to eventually be members of their *people* in good standing.

She models for her children what a woman of her *people* is and teaches her daughters what a woman of her *people* must know. He models for his children what a man of his *people* is and teaches his sons what they must know.

His natural inclination is to provide for and protect his woman and their offspring, and in doing so he feeds his masculinity. Her natural inclination is to make a home for and care for her man and their offspring, and in doing so she feeds her femininity. Their mutual gratitude and adoration feeds their respective masculinity and femininity. This masculinity and femininity are critical for the survival and progress of humanity because they are key ingredients in producing the masculine men and the feminine women that can stand in the face of hardship and produce children capable of the same endurance.

This is the *natural-rational-family* born of our *healthy-nature*. This fruitful union is timeless and universal because it is born of our respective human natures. This union *is* marriage and marriage *is* this union and not subject to redefinition.

◆ ◆ ◆

When the man and the woman together submit to what nature has equipped and inclined them to do, the family is formed and made strong, and *via* intermarriage with other strong families new families are formed that perpetuate themselves and their *people*.

This has been the nature of things since man cut with flint and woman cooked over a dung fire. It remains the nature of things today, and will be the nature of things when woman makes her home in zero gravity orbiting Jupiter or on worlds and moons as far as man should ever travel.

However, man has another side of his nature that at all times wars with his *healthy-nature*, which combined with the *Scientific Revolution*, and its twins the *Enlightenment* and the *Industrial Revolution*, have produced a time of unprecedented challenge to man, woman, marriage, the family and the *peoples* they make.

VII. THE UNCHANGING
FAMILY IN CHANGING TIMES

WERE ALIENS TO COME to Earth to explore it and observe its life they would find it teeming with human beings. Individual people. Billions of them. It would be manifest to them that these humans are far and away the dominant life form—hogs, though the most delicious, being a distant second.

They would also notice that these humans continuously compete with each other for survival, prosperity, and control, and that this competition is often ferocious.

It would be equally manifest that these humans live in small family groups and that their loyalty to their own family members exceeds by a long shot their loyalty to other humans or the other manifest human groups they form.

Families are simply everywhere. Go into any restaurant on a Saturday evening and look at the tables. There they'll be. Look inside vehicles on the interstate. Drive around the countryside or through any town. What do you see? House after house, apartment after apartment, the vast majority occupied by the ubiquitous human family. Few of us live alone and of those that do only a handful can lament that no one would give them a kidney if they were in need of one.

Families have always been and will forever be the fount and fruit of human life, and the centrality of the home in human affairs has not changed since Eve roasted her first rabbit outside the garden of Eden while Adam whittled the sharp end of his spear. And it never will change.

But times do change and people are as unique as the *peoples* they form, including *We the People,* and indeed something unprecedented has happened that has affected the home in this era on a scale that the historian would only have expected to see during times of great loss of life such as war, famine, or pestilence.

♦ ♦ ♦

It is human nature to imagine that the way things are is the way they've always been, and once the new thing is established to forget how it was before. It is hard to imagine life before the invention of the wheel or the adoption of Arabic numerals, or life before the great mass of women did not work at home. But it was not long ago when you could knock on the door of any dwelling (whether brick, adobe, buffalo hide, thatch, or whatnot) and expect to find the lady of the house there. Thus it had been from time immemorial.

This suddenly is no longer the case today. What has changed? How is it that woman working outside the home is now the norm, or at least very common? And why didn't this happen a long time ago?

It is easy for us to think of our ancestors as unenlightened or fools, and maybe at times they were, but to think thus without any understanding of the times and circumstances in which they lived is to deny ourselves the benefit of understanding our roots, and what has occurred that has made things as they are today.

Feminism is guilty of this type of thinking on a grand scale and has perpetrated one of the great deceits of our times, namely, by declaring that women throughout the ages have been kept at home against their will by men. While men were free to come and go as they wished and otherwise make *his-story*, women were forced to toil in drudgery at home, which was in fact a prison. With the coming of *Feminism*, indeed just an idea, women were at last liberated from the house and the bonds of marriage, which was in fact slavery.

The truth is as different as it is obvious. Before the modern age, mothers, wives, and daughters were to be found at home for the same reason that spacemen are found in spaceships. It was impossible for her to survive outside of it. In fact, it was impossible for mankind to survive if she was not there.

The modern woman is untethered from home life to the extent that she is because of an unprecedented window of time that has opened to allow it, not because some 19th-century feminist had an *aha!* moment and thought to herself, "This ain't fair, and we ain't gonna take it no more!" No, she said, "This ain't fair, and we don't *have* to take it no more!"

It is likely that Adam looked up at the birds flying through the air, and beholding them thought, "Dang! That'd be handy!" And ever since his sons and daughters have imagined what it would be like to fly. But they didn't fly until the Wright brothers figured out how. Likewise, Eve doubtless envied Adam's great strength and freedom of movement and wished she could wear the loincloth of the family but alas she could not. She could only lament the endless work to be done in the cave, while Cain and Abel and the other rugrats were ever underfoot crying, "Mama, mama, mama, mama, mama, mama," until she wanted to scream or in fact did. Then there was Adam, still brooding over the incident with the serpent, forever wandering off into the woods to do Lord knows what, and only coming home to eat and sleep, or when he had an itch he needed her to scratch. When he did talk it was all yap, yap, yap about that amazing shot he took or getting chased by a boar or wondering how he was going to get across that river, all the while ignoring, tripping over, and sometimes just flat-out stepping on her feelings. Then there was that terrible evening when he had gotten too deep into that infernal honey-water concoction of his and started carrying on about the apple tree again, and she told him with a little heat that she thought perhaps it was the serpent that had given him the honey-water recipe, and come to think of it, if he was so smart how is it that he was so easily duped into eating that cussed apple? "Do you think I like living out here in this dank, smelly..." But her speech ended there when his open palm hit her so hard across the face that it sent her one end over the other across the fire and onto the dirt floor. And before she could come to her senses again he had leapt for her, taken her furiously by the arms, jerked her up off her feet, and shaking her violently, screamed the most hateful things in her face before flinging her back onto the dirt floor where she lay limp. For a moment. Then her tears came, and those tears turned into sobs, her sobs into wracking sobs, and her wracking sobs into convulsive wailing. She had always known that he didn't love her, and rising up on one hand through snot, tears, and choked howls she told him as much and several other choice things besides before collapsing upon her face to writhe in the dirt in a paroxysm of hurt, loneliness, and despair.

It was then Adam's turn to be shocked for he had not seen her like this since little Lucy had died within a day of becoming ill. My God, what pain!

And the fury left him as quickly as it had come, to be replaced with that cursed guilt and shame. He leapt to her again but this time to comfort her, and for half an hour to apologize and beg forgiveness before she would finally turn her dirty face to him again and let him rest his hands upon her and eventually be drawn into his embrace. And there they huddled for a long spell and really talked, and he listened to her and seemed really interested in her feelings, which was nice, and finally with the fire but meekly glowing they made up properly, which was even nicer, and for days afterwards he was more thoughtful than he normally was, and that was nice too.

Then a week later he didn't come back to the cave at all one night, and sure enough Abel got sick and kept her up all night with a cough, and when Adam did finally come home around noon the next day he just wolfed down the rabbit stew she served up without even commenting on why it tasted different, much less compliment her on it—for it did indeed taste better because she had put a new little leaf she'd found in it because she thought he would like it, much like she thought he would like the way she had done up her hair, which he had not noticed either before collapsing on the fresh grass mat that she had made up for him the day before because heaven forbid he sleep on dry grass! And now she had to herd the kids out to keep from waking him, and if they didn't gather up the pecans now the squirrels would get them all!

And looking up at the blue sky through the trees she cussed her lot in life but not God, though briefly tempted, for she both feared and trusted Him. No, she would just endure. Endure because she could not change her lot in life anymore than Adam could fly.

Since the beginning of time women's grievances against men have been understandable because, like their father Adam, men are just thoughtless jerks. Or they can be. But do we imagine that the lives of the sons of Adam have always been peachy? Yes, the daughters of Eve were stuck in the cave and under a husband's thumb, but he was likewise bound to the fields and woods that sustained them and was himself under the thumb of some kind of *Captain* that he had to follow, and who was not always just or generous. No, she wasn't able to roam about at will, but neither was he able to roam safely outside the territory controlled by his *Captain* and the other men of the tribe. Yes, her life was always full of toil, and birthing babies took its toll, but his life was also hard, and it has always been the sons of Adam whose bodies were broken by years of the labor and injuries that have always been part and parcel of the life of the man who went outside to subdue the Earth, and it has been their bodies that have lain in great heaps on the battlefield.

And do we imagine that Eve didn't get under Adam's skin now and then, driving him to the proverbial rooftop to cuss that woman and finger that nasty scar on his side where he used to have a perfectly good rib? And even though she wasn't able to beat him into submission, do we imagine she didn't possess her own more subtle means of bending him to her will? Whether sweetness or sensuality, wiles or cunning, nature has granted to woman power more than sufficient to bend or break the will of the stronger sex and to leave him either blessed or wretched. The Jezebels, Delilahs, Livias, Fredegunds, and assorted shrews of *history* have always given the wife-beaters a run for their money in the misery they have produced.

Life is hard under the best of circumstances for woman *and* man. It is made hard in large part because men and women are both just awful, or at least awfully self-centered. And where two or more of them come together, trouble and drama will likely follow. In no relationship is this more true than within the bond of marriage. Yet men and women have always paired off; women have always made a home, men have always lived in those homes, and women have always been stuck there.

At least until recently, that is. Beginning in the 19th century *Feminism* reared its ugly head, and on the grounds of timeless and universal female suffering, as well as the novelties of *Freedom* and *Equality,* has waged relentless war against the timeless order of things. Well, I call it war. *Feminism* has certainly challenged with marvelous success the aboriginal paradigm of the wife at home doing woman's work, subordinate to her husband who works away from home doing man's work. The human wreckage produced as a result is manifest, so yes, I call it war.

But why didn't *Feminism* emerge in 1500 AD or 1500 BC or right out of the gate? Why didn't our first mother stand up and say...

"Adam!"

"Yeah babe."

"I'm going out to hunt today. You stay home and take care of the kids. You need to finish turning the clay pots and then go out to gather the blackberries. Then I want you to take those rabbit skins there and make yourself a nice little pair of briefs—you know, the kind that ride a little high. What's that? No! I told you, we're not moving to another cave! Why? Because I said so! Oh really? In that case you're just out of luck! Fine, leave then! What do I care. Hey, maybe this thing has run its course. Let's get a divorce."

But Eve didn't kick Adam out of the cave or hit the road and for ages on end neither did her daughters. Why not?

Because they couldn't.

Feminism was able to raise its head only in the 19th century, ask the questions it asked, challenge the ancient order of things and demand changes to it because the notions of *Individuality, Freedom* and *Equality* born of the *Enlightenment* had by then taken deep root in our collective mind. Indeed, men of the working classes had begun to achieve a greater degree of *Freedom* and *Equality,* so why shouldn't women as well? But equally necessary for women's liberation was the extraordinary wealth and technological advances being produced by the *Industrial Revolution* that had made *men's* liberation possible. The ideas of *Individuality, Freedom,* and *Equality* were one thing; the ability to put them into practice was something altogether different, and the ancient agrarian order had always presented too many obstacles to realize such fine notions.

But working hand in hand, the *Enlightenment* and the *Industrial Revolution* appeared to solve the problem.

In the last three hundred years industrialism has driven agrarianism before it just as thousands of years ago agrarianism drove nomadism before it. With astonishing quickness we have seen the great bulk of mankind move from agricultural to industrial work—the most manifest sign of this change being the mass movement of people from the countryside to the city. From field, pasture, and shop man would move to factory and office. And this move into the cities would see the near disintegration of ancient agrarian orders and traditions, and a greatly reduced dependence upon extended family. Combined with an attendant explosion of wealth, the stage had finally been set for both man and woman to ask heretofore unutterable questions, and, in the case of the latter, *Feminism* would in time produce the ultimate *argumentum ad absurdum*, namely, the debate at hand.

But as radical as these changes were, man's living paradigm remained at its most fundamental level unchanged. He still left home to work and returned at the end of the day. He still built the things that were built and if he didn't grow nearly as much as he used to he still made stuff. Though he was no longer responsible to an extended family, clan, or tribe, he still formed and joined groups of all kinds. Though he no longer answered to a chief or lord and now viewed himself as free and the equal of all men, he would still answer to some man somewhere and he certainly still fought in the endless wars that are an unavoidable part of human life. Lastly, it did not occur to him that he wasn't the head of his house or that it was not his responsibility to provide for and protect his family.

But the impact that the *Enlightenment* and the *Industrial Revolution* have had upon woman and through her, man, is unprecedented in human history.

Up until the modern era any discussion of man or mankind presupposed woman, and when a man was spoken of individually it was understood that he was part of a whole, the other part being a woman—his better half, his wife, the mother of his children, the keeper of his home.

Between the elements, beasts, and most importantly the violence of men, the natural world had always been a dangerous place for both man and woman, but especially for woman. Her strength, toughness, and endurance, after all, cannot compare to his, and she is further weakened and made less mobile by pregnancy and the needs of her infants and young children.

In a harsh and dangerous world she made the home that was her refuge, and the presence of her man made it her castle. The homes of her *people* made up the human hive that protected them all.

Besides offering security her home was also a place of production where she did those things that by nature she was inclined to do, that she alone could or would do, and that had to be done if her family, which she loved more than her own life, was to survive.

She had always been at home because she had no choice. Home was where she was safe and where her work was.

The coming of the the *Industrial Revolution* and the *Enlightenment* and its offspring *Feminism* has not eliminated this paradigm of home as refuge and place of production, but they have radically reduced its sphere. Since the 18th century institutions emerged that removed the great bulk of a woman's work from the home, and technology has greatly reduced what remained. Additionally, institutions emerged that likewise reduced her dependence upon her husband's natural role as her provider and protector. Marriage itself has been greatly weakened by ideas that would tempt or cause her to view herself as an individual first, a daughter, wife, and mother second; and tempt her to place the solidarity of all womankind above that of her family and *people.*

Eventually, she would hear voices that would actively encourage her to despise homemaking, motherhood, and especially marriage, and voices encouraging her to pursue—well—other things.

◆ ◆ ◆

One of Feminism's great propaganda successes has been its ability to associate the idea of women's work with work outside the home, and by this we do not mean work in the garden or downstairs in the family's place of business. Rather, we mean work geographically removed from her home, usually in factories, offices, or service businesses.

This woman is the *working woman.*

"Do you work?" is a loaded question that all adult women are asked today. A positive reply means yes, she does work away from home at some place of business. A subtle respect is associated with a positive answer, and a subtle contempt with a negative answer. She doesn't work. She's just a *housewife* and who can respect that?

For generations *Feminism* has intentionally and deceitfully cultivated the image of the professional (or at least independent) woman that works outside the home. *Feminism* liberates woman from the prison of home to go forth into the world to do meaningful and respectable work—like man had always been able to do. This image is implicitly contrasted with that of the pajama'd housewife lounging around the house all day watching soap operas until her lord and master returns in the evening, at which point she becomes cook, scullery maid, and sex slave.

But such caricatures ignore historical reality. Aside from a few pampered women of the affluent classes, pre-modern women always worked.

It was she that gathered, grew, raised, preserved and prepared our food.

It was she that tanned the hides or made the cloth that kept us clothed.

It was she that cared for our young and elderly.

It was she that doctored the sick and injured.

It was she that educated her young, especially her daughters, and taught them what they needed to know in order to survive in their times amongst their *people.*

It was she that made the things she needed or could trade.

Last but not least, it was she that expedited her husband's work.

The home had always been a place of production and a place where things that had to be done were done. And these things were produced and done by her because it was her inborn nature to do them, not because Eve lost a bet to Adam, or because he beat her into submission, or because their sons and daughters had just kept up the tradition.

Woman is a natural born worker, multi-tasker, and manager. Like making a home and making babies, it's in her DNA. And until the modern age, and in living memory in West Tennessee, she exercised these inclinations at home because that is where the work was and the only place it could be.

For the first time in *history* this is no longer the case.

She no longer produces and prepares food, but instead goes to the grocery store or out to eat food mass-produced—who knows where—by *Big Food.*

She no longer spins, weaves and sews to keep us clothed, but instead goes to Walmart (or Target if she is a good *Progressive*) to buy off-the-rack clothing mass-produced in textile mills in Timbuktu.

She no longer cares for the young or the elderly. Instead, we send our young to daycare and our elderly to nursing homes.

She no longer takes care of the sick and injured. Instead, we send them to doctors' offices and hospitals.

She no longer educates our young. Instead, we send them to school.

She no longer makes the things she needs, but drives a great distance in a vehicle powered by a combustion engine on blacktop roads to Walmart to buy what she needs with cash that she earned making stuff in a factory or doing stuff in an office.

As for the work that remains in the modern home, technology has simplified it across the board. If she needs light, she flips a switch. If she needs to cook, she turns a dial on the oven. To heat food she pops it in the microwave. To preserve it she puts it in the freezer. If she is cold she turns up the thermostat.

If we keep our bodies and clothes cleaner today than did our ancestors, it is not just because the relationship between health and cleanliness was not always understood. It is also because our ancestors didn't have indoor plumbing and gas water heaters, or washers and dryers.

Woman hasn't left home to work for the first time in history because she was liberated by an idea, but because her work at home has dried up, and necessity and her natural inclination to work have compelled her to follow her husband to seek employment away from home in offices, factories, and elsewhere.

Like her husband she would now specialize in these new places of employment. Instead of wearing many hats, she would become a nurse, teacher, secretary, or line worker. Just as it did first for her husband, industrialization would eventually explode the number of occupations available to her.

These new institutions, technological innovations, and especially cash-paying jobs, besides having liberated her from home for the first time in *history*, have also liberated her from her natural dependence upon the men of her family who protected and provided for her, and from a husband in particular. But there would be many more supports in this unprecedented liberation.

Life insurance, health insurance, liability insurance, unemployment insurance, auto insurance, workman's compensation, deposit insurance, property insurance, etc., would protect her against the vagaries of life.

The welfare state would provide for her and her children when she is young.

Pensions, IRAs, Social Security, and Medicare would provide for her when she is old.

The modern police state with its numberless tentacles and vast prison network would keep dangerous men at bay, and a call to 911 would see dangerous men with badges fly to her aid.

But to fully liberate woman, *Feminism* needed more than just the material means. It needed the force of law.

Under the banners of *Freedom* and *Equality* any prohibition of her owning property had to be eliminated.

Linked to the new police state she would gain legal protection against her own husband. The police would now enter directly into her home and clasp her husband in irons if he raised his hands against her.

The courts would now grant her a divorce for any reason whatsoever and forbid any attempt on her husband's part to lay claim to her.

Lastly, he would be forced by law to continue supporting her via child support and alimony.

Concurrently with and a driving force behind all these sweeping changes, *Feminism* worked tirelessly to persuade us to view woman as an individual first and foremost, and her liberation from home and husband as absolute goods and necessities.

Then to top it all off *Feminism* spawned *Sexual Liberation,* which would completely overturn both a rational view of sex and centuries of Christian influence that had placed some restraint on this most powerful of human imperatives. Sex was not only *not* to be restricted to marriage, it was now to be actively encouraged outside it, no holds barred, without shame or stigma. The development of modern contraception would free woman from the fear of unwanted pregnancy, and access to a safe and legal abortion would render any such pregnancy a mere inconvenience rather than a lifetime commitment.

Indeed, by the 1970's the *Enlightenment* and the *Industrial Revolution* had created economic, technological, legal, moral, and cultural circumstances which for the first time in history would allow women to work outside the home, to live alone, and to copulate at will. They could have children or not, within or without the bond marriage, and abandon their husbands or children in order to improve their lives.

Incidentally, if woman would be liberated, so would man.

All this new freedom and independence was great, except that the result of both man and woman thus being able to do what they wanted— or in the case of woman, being compelled to leave home to work—has been in the last fifty years an explosion of manless homes, broken homes, no-homes, and illegitimacy.

Feminism pats itself on the back for a job well done because it holds *Freedom* and *Equality* of woman as sovereign gods... I mean goods, does not accept the human wreckage that it has created as such, and does not accept the following fundamental truths of human nature: the family is the building block of humanity; a man plus a woman make a family; man is the natural head of the family and woman is inclined to follow her man; children need both of their biological parents; they all need home, and woman makes that home. She is in fact that home.

◆ ◆ ◆

What would be left if everything that was built by the hands of man were to vanish? A great deal of wilderness, I imagine. This is because for humans it is the male of our kind that builds. He does so because it is his natural inclination. Woman does not do much building because it is not hers.

What would happen if he were to stop building? It would not be long before the *people* would suffer greatly, because we are sustained and protected by all that he builds.

45

What would happen were he to stop producing? It would not be long before the *people* would begin to suffer great want.

What would happen were he to stop fighting? It would not be long before we would be servants of other men that would fight.

If man stops doing what he does by nature, i.e., building, producing, and fighting, then society will collapse under the external threats of the elements and our enemies.

But what would happen if woman stopped doing what by nature she is inclined to do?

There would be no home, no comfort, and no purpose. All would be chaos, confusion, and disorder. It would be a hive without a queen, an eternal stag party. Society would collapse from within and man would be defenseless against it. He would be all today and no tomorrow.

Collectively man continues to succeed and fail as he always has because there is no *Men's Liberation* movement encouraging him to do otherwise. He hasn't stopped building the structures in which we dwell or the roads that connect our cities, or producing the stuff that we need and want. Neither has he stopped defending us. In our military and police forces, and in his own home lying at his woman's side—man, as soldier, cop, and king of his castle—keeps at bay enemies without and within. With woman as his central load-bearing pillar he is superbly constituted to protect against the external threats of man and nature. Ensconced in his castle, she gives him ultimate purpose here on Earth. She is the cement in his lime, his North Star, his magnetic pole.

But for generations the internal rebellion that is American *Feminism* has worked tirelessly to wreck this central pillar and compass of human life and the cultural rot that it has produced is everywhere around us.

♦ ♦ ♦

To be fair though, in spite of the manifest cultural decay in the United States caused by *Feminism*, our society has not collapsed from within because the majority of women, despite paying lip service to *Feminism*, still live rational lives that their mothers going all the way back to Eden would recognize. Yes many are compelled or choose to work outside the home, but it is still she that manages it. Many women do not spend as much time with their young as their foremothers, but they still love them and tend to their needs as only mothers can. *Mother's love* is after all a universally understood concept.

Many women still remain faithful to husbands, even flawed ones, and are their true support. Many women still postpone childbirth until marriage. Many women exercise self-discipline and restraint, denying themselves and living for their families.

Many women still embrace their feminine *healthy-nature* and there are still mothers everywhere producing good kids that grow up to be decent men and women who make positive contributions to their *people.*

But in the last fifty years the sheer number of children whose mothers are away from home while they are young has collectively weakened the influence of mothers and home life. That the new post-*Industrial Revolution* economy has made this unavoidable is the conundrum of the age, but how could it possibly be good?

But what makes *Feminism* a positive evil is that it has made a less-than-ideal situation vastly worse by not only promoting the separation of woman from home, but by denigrating homemaking itself. Furthermore, it has striven to eliminate all distinctions between man and woman, has rejected the *natural-rational-family* that sees man and woman as a disciplined whole, has promoted promiscuity and divorce, and has encouraged woman to live first and foremost for herself, and not for her husband and children.

Feminism has succeeded because of the secure and prosperous times in which we live, and because it has been championed by an American media dominated by *Progressives*—a propaganda machine that would have made Joseph Goebbels green with envy.

The result of its success has been an unprecedented catastrophe. The evidence of this cultural rot is everywhere, including out here in the countryside of West Tennessee, and *Feminism* is largely to blame for convincing us to reject our *healthy-nature* by playing upon our darkest and weakest instincts as human beings.

VIII. HUMAN DUALITY
How the *Bad Idea* That is *Feminism* is
Spawned by Our *Unhealthy-Nature*

YES, THE *ENLIGHTENMENT* and *Industrial Revolution* have had a profound and unprecedented impact on the way we live and on the home in particular, but what they haven't done is change human nature. We humans have one and it cannot be escaped. Yes, it is clearly in man's nature to produce, build, and protect; and yes, it is a woman's nature to make a home for a man, herself, and their offspring. Together they make the *natural-rational-family*, which is the ideal.

But if this division of labor is natural, then why is it that throughout history it has so often failed to achieve the ideal? Indeed since the beginning of time the reality has been less than ideal and the result has always been homes and people ranging from the quirky to the hopelessly dysfunctional.

In a nutshell the reason is because life is full of contingencies and man is imperfect.

◆ ◆ ◆

Men and women will forever be subject to accidents and events beyond their control that impede their ability to perform their natural roles of husband-father and wife-mother.

Sometimes we fail to fulfill our nature because we are simply not intellectually or physically up to the challenge that life has dealt us, or that we have dealt ourselves because of foolish decisions or honest mistakes.

Troubles and calamities are the nature of things in human life. Nature, accidents, events beyond our control, folly, and weakness will forever leave men and women sitting on the edge of their beds weeping in their hands.

As a result of life's troubles and calamities we all suffer pain, and sometimes this pain translates into what we call *baggage* or *dysfunction*, that state of mind that impedes contentment or that inclines us to behave in such a way that needlessly injures others or our own best interest.

Troubles and calamities beget dysfunction, which in turn begets more troubles and calamities—a cycle that is self-perpetuating from generation to generation.

As an extreme example, consider the civil war that has been raging in Syria. By death and severe wounds, that war has prevented untold numbers of Syrian men and women from performing their roles as husbands and fathers and wives and mothers. The rest have been marked to one degree or another and many have been deeply scarred or broken. This latter group will for the remainder of their lives (to one degree or another) wound everyone they are ever close to, especially their own children.

Troubles, the pain they cause, and the baggage caused by the pain are all unavoidable. It's part of life. By wisdom, will, and strength (or by God's grace if you believe in Him) all we can hope to do is minimize them until such time as we cross the Jordan, the Styx, or just rot.

However, there is one cause of trouble and pain that dwarfs all the others combined in the amount of pain caused. I speak of that which inclines us to behave in a way which pleases or benefits us but hurts our own selves or others unnecessarily. It is that aspect of our nature that is most like that of animals. It is the fount of weakness, thoughtlessness, sloth, neglect, lies, theft, meanness, hate, vengeance, and cruelty. It is what we recognize as evil and has been aptly called *the flesh*, always and forever the greatest source of human misery. It is an inescapable inheritance of the first humans, and has always produced most of the troubles and calamities that have befallen mankind and thus most of the dysfunctions that we bear and in turn pass on to our children, even as they were passed on to us.

I will call this fleshly nature which inclines us to do that which is desirable but unwise or unprofitable our *unhealthy-nature* and contrast it with our *healthy-nature*. I say *nature* because they are both equally parts of our human nature which we can no more change or escape than the birds can theirs.

These two natures, equally human, wage relentless war on each other, and this conflict is the principal source of the history of man. After all, remove hate and love, or pride and humility, or gluttony and temperance, or sloth and industry from our nature and history would be considerably less interesting.

In a nutshell, our *healthy-nature* produces strength; our *unhealthy-nature* produces weakness.

Our *healthy-nature* inclines us to do that which is wise, which in turn yields the best fruit. I will declare this best fruit to be peace of mind, or the well-being and survival of our children and our *people*. It produces *quality of life*.

It is our *healthy-nature* that inclines us to love and sacrifice for others and to enjoy that love and sacrifice returned in kind. This love that we share is imitated by those that receive it (just as dysfunction is) and is passed on to others, binding us, making us stronger together.

It is our *healthy-nature* that inclines us to do that work that strengthens our mind and body, satisfies the soul, and that sustains ourselves, our families, and our *people*.

It is our *healthy-nature* that at all times makes war on our *unhealthy-nature* by inclining us to do that which we do not want to do but which benefits us or our fellow man, and inclines us not to do that which we want to do but which would be injurious to ourselves or our fellow man.

Monogamy vs. promiscuity is a prime example of this conflict.

Man thinks about woman a lot and pretty much sizes up every woman that comes into view. It's not that he wants every woman that he sees or that there is a latent porn director in every man, but the idea of sex does flit across the mind of even the most chaste and self-disciplined of men.

So does this make men like the bull in the pasture who is ready to go in a moment's notice as soon as any heifer gives the signal? Not quite, because there is another side to man's human nature that wants something every bit as much as he wants sex. That thing is a home that only a woman can make. That woman is called *wife*.

Man cannot have both, because woman has her own nature. She is not nearly so tempted by her *unhealthy-nature* to take multiple mates, but is rather inclined by her *healthy-nature* to take one and then become intensely devoted to him in thought and deed. She is further inclined to give him the home and adoration that he craves. However, by her nature she is fiercely jealous and will not share her mate with any other woman, and should her man spread his affections elsewhere he can wound her so deeply that she will be unable to adore him as he would wish. While she may do his laundry and fry his eggs, her heart will not be in it—without which his home is just a box and she a maid.

Man is just as inclined to mate for life as he is to rut like a stag. Do we say, then, that he is promiscuous or monogamous by nature?

Well, he is both. His *healthy-nature* inclines him towards monogamy, his *unhealthy-nature* towards promiscuity, and we know which nature is healthy and which is not by their fruit.

Monogamy contributes mightily to the contentment and security of the wife, who in turn contributes mightily to the serenity and contentment of the home, which contributes mightily to the well-being of the husband and the children. As they say, *happy wife, happy life!* A contented wife and a stable home support him in his labors and thus serve his family's prosperity. Furthermore, lifetime monogamy avoids one of the greatest destroyers of wealth: divorce. Her monogamy means that her children are his, which is by nature of great importance to man. His concern about his paternity inclines him to be, unlike the bull in the pasture, more responsible for his own. Monogamy means that children are raised by both of their biological parents, which is best for them, exceptions notwithstanding. Monogamous couples don't contract STDs or produce unwanted children who themselves are laden with the dysfunctions associated with not being raised by one's biological parents.

But promiscuity simply leads to all manner of troubles, calamities, hurt, baggage, and dysfunction.

Both monogamy and promiscuity are natural, but one leads to the good, and the other to the bad.

Work or restraint are the hallmarks of our *healthy-nature*, and satisfaction and increase is its fruit. Ease or pleasure are the hallmarks of our *unhealthy-nature*, and tears and decay is its fruit. Just compare the hardworking, active man who eats good food and enjoys his pleasures in moderation to the video-game-warrior-porn-addict who works only enough to support his pleasures and lives on root beer floats, pizza, and potato chips.

How we answer the call of these two natures dramatically impacts our own peace of mind and physical health as well as those whom we influence, and thus the survivability and prosperity of ourselves and our offspring.

And also our *people*.

The strengths and dysfunctions that are bequeathed to us by those who raise us, and which we pass on to our children and others that we come into contact with, can become so widespread amongst a *people* that they can become their cultural markers. In the South we find good examples of this. Southerners are well known for being friendly and quick to forgive, but it has also been said that they don't always tell you what they are really thinking and hold silent grudges. I believe there is truth in these stereotypes. If true, it is because of the timeless conflict of the two natures of the first humans, but more specifically because of the immediate ancestors of Southerners that came here to the Land of Cotton (a.k.a. Dixie or God's Country), the centuries that their descendents have mingled together, their shared experiences, and now memories of that time. Just as the Mexican has emerged in Mexico since 1521, the Southerner has emerged in the South since 1607.

Just like individuals, entire *peoples* can be shaped by the movements and great events of their times that are outside their collective control. Ideas, war, famine, and pestilence will alter their character as well as good or terrible leaders who lead them through the shoals or dash them against the rocks.

The upheavals of great events can strengthen *peoples*, but they can also weaken, bend or break them. Syrians may be living through such a period now. Mexicans did between 1910 and 1920, and Southerners between 1861 and 1876. The list could go on and on and on.

It does not matter that the mass of these *peoples* may have been living quiet, peaceful lives, filled with work and bridled by self-restraint. Living well can offer little protection when the interests of entire *peoples* clash or when a rat scurries off a Venetian galley carrying a plague-infested flea that will kill upwards of half the population of Europe.

Some bullets can't be dodged.

But others are self-inflicted, the result of a *people* embracing a *bad idea* or making a bad decision, only to be exposed as such in hindsight and sometimes by historians performing post mortems.

Regardless, whether a *people* are swept up in events beyond their control or whether they embrace a *bad idea*, sometimes the outcome is just more than they can bear and they are never the same, and just as individuals, will bear scars for generations in the form of collective quirks, baggage, and dysfunctions.

The American *people* are not immune to this and neither is this age.

But before we can assess the strengths and weaknesses of the American *people* it must be noted that ours is not a monolithic culture, not in the way the Polish or Japanese cultures are with their homogenous *peoples* still occupying the same land their ancestors occupied in prehistoric times and still speaking the same language. No, the American *people* are too young, have been drawn from too many disparate *peoples*, and occupy too large a space for us to point to a single American type.

So what do we have in common that will permit us to even refer to an American *people* or an American culture? The US is a nation encompassing half the North American continent, and includes a far flung arctic state and an even farther flung Pacific island state. Given such expanse and diversity the only commonalities that Americans share are the English language, a legal system, and a general respect for freedom and equality, though there is little agreement on the nature of these last two points.

While it is true that all people are born equal in the sight of God (if you believe in Him), and while it may be true that they should be equal in the eyes of the law, men are from birth manifestly not equal (or the same, if you prefer) and these inequalities or differences only increase over time. We can hardly be born above or below our parents, and we will be poured into the mold that is our family and the *people* of which that family is a part and be shaped by the times and place in which we grow up. All of this produces endless inequality. Not only are we *not* the same, some men are decidedly better than others, and this difference is directly proportional to the degree in which they, their parents, and their *people* have embraced their *healthy-nature* vs. their *unhealthy-nature*, or *good ideas* vs. *bad ideas.*

Just as some men are better than others, some *peoples,* or the *sub-peoples* or sub-cultures that make up a greater *people,* are better than others and for the same reason. Consider the subcultures of collegiate wrestling and that of the college party culture. College wrestlers must embrace the physical and mental discipline of their *healthy-natures* in order to succeed on the mat. The result is not just a young man in superb physical condition but one who also possesses a high degree of confidence in himself. He possesses the ability to do the hard thing and the courage to face the dangerous thing, both attributes that will serve him for the remainder of his life, no matter what may ambush him along the way.

But the young man that gives way to his *unhealthy-nature* and spends four years buzzed, drunk, and hammered, hoot'n and holler'n with his boon companions, chasing drunk college girls and a 2.5 GPA only weakens his body and mind, which limits the paths he may choose and makes every path harder, and decreases his chances of surviving when the zombies come. And in one form or another the zombies always come.

The culture of collegiate wrestling is simply better than that of the collegiate party culture. By their fruits ye shall know them.

Likewise we know the strengths and weaknesses of our American culture by its fruit and the fruit of its many *peoples* and subcultures. Its strengths we will readily acknowledge, but its weaknesses we are not so quick to concede. But who will deny that American culture is flawed or that it has produced some dysfunction?

Of course it is and of course it has! And I posit that when historians look back upon contemporary American culture from a dispassionate vantage point there will be general agreement that one of the greatest cultural markers of our age and the greatest producer of dysfunction in it shall have been the manifest breakdown of the *natural-rational-family.*

♦ ♦ ♦

We humans possess a *healthy-nature,* and this nature inclines us to produce instinctively and forever the *natural-rational-family.* Inasmuch as any ideal can be reduced to a formula, I posit as sovereign truth and offer no evidence beyond what common sense observation can provide, that the ideal family, born of our *healthy-nature,* consists of:

- A masculine man and a feminine woman united for life,
- in mutual respect,
- in a sexually exclusive union,
- the man leading,
- the woman following,
- the woman providing them and their offspring a home,
- the man protecting and providing for them and their home,
- both committed to the well-being and development of their children,
- both emotionally and physically affectionate towards each other and their children.

This family is most likely to produce the well being and survival of children and by extension the *people.* This family is the greenhouse that is most likely to produce the young man or woman that will withstand the vagaries of life, and make, support, and protect what is good, passing on the strengths of the *people* to the next generation. It is least likely to produce the young man or woman that is laden with baggage or dysfunctions.

That this ideal is rarely achieved does not justify our chucking it out the window. It is in fact broadly achieved by many, and many more over the course of a lifetime will make progress towards it. Indeed, the strength of any given *people* is a product of the degree to which its constituent families are able to embrace the ideal of the *natural-rational-ramily.* The further removed a *people's* families are from this ideal, the weaker the *people.*

Just imagine the kind of men and women that would be produced by the anti-ideal family. This anti-ideal family is as rare as the ideal family, but semi-dysfunctional and dysfunctional families are not uncommon in any *people* or any age and they have always been and will forever be the principal fount of emotional pain, baggage, dysfunction, sexual deviancy, and criminality.

But what is remarkable in this era is the number of dysfunctional and semi-dysfunctional Americans that have been created since World War II and the Cultural Revolution of the 1960s in particular—remarkable, because this has been a period of extraordinary peace, prosperity and so-called *Progress*. Shouldn't this have created a more balanced and contented American on average? But instead we are seeing dysfunctional Americans on a scale that one would expect to find only in times of great loss of life such as war, famine, or pestilence.

This has come about not because we have been swept up in events beyond our control, but because *We the People* have embraced *bad ideas*, namely elevating *Freedom* and *Equality* to sovereign status, making them the foundation upon which we have ordered American society rather than important and honored elements of it. As a consequence we have embraced the worst conceivable *bad idea: Feminism*.

The goal of *Feminism* was never limited to gaining for woman political power, equality before the law, protection against abusive husbands, or equal pay for equal work, but rather the complete elimination of all distinctions between man and woman, at any cost, in order to achieve for woman *Freedom from* and *Equality with* man. Indeed to make them not just *equal* but *the same*.

Feminism would have us believe that man and woman are the same or ought to be the same. That their manifest behavioral differences are cultural constructions supported by millennia of tradition, rather than born of our DNA, or that their physical differences should be of no consequence or be made of no consequence.

It would furthermore see woman liberated from man.

Feminism does not view man as a half with whom woman is to be reconciled in order to make a complete, purposeful whole, but rather as the enemy to be defeated, or from whom she must be liberated.

By means of a *Progressive* propaganda (print, radio, cinema, television, and now the Internet) and educational establishment *Feminism* has been able, because of the very peace and prosperity that resulted from our victory in the Second World War, to preach that human nature is not what it appears to be and ask heretofore unaskable questions. *Feminism* has successfully tempted both man and woman to pursue their *unhealthy-natures* by seeking that which they have alike wanted since the beginning of time—namely, to have their own way.

But all of this pretending that we can be something we cannot be has only created an illusion that hides the deep cultural rot that the pretending itself has produced, rot that is even now drawing to a close this prosperous age that allowed us to do the pretending in the first place.

By its fruit we may know *Feminism*.

The Fruit of Feminism

What the preaching of *Feminism* has done is to produce on a massive scale generations of confused and ill-prepared boys and girls that have grown up to become confused and ill-prepared adults, ignorant of their fundamental natures as man and woman and thus incapable of producing sound families.

War Between the Sexes

The union between man and woman is universal, timeless, and fruitful because it is born of our human nature. This union is not always blissful, which is equally born of our nature. However, this natural conflict can result in a much stronger union that is better able to perform its mission. *Feminism*, by stoking woman's timeless and understandable grievances against man, by inflaming her natural envy at the power and status of man in the natural order of things, and by magnifying womankind's natural sense of solidarity, has started and sustained in this prosperous and peaceful age a war between the sexes that has produced a profound, enduring, and highly destructive disharmony between American men and women.

Divorce and Illegitimacy

Healthy cultures regulate procreation because it is a sovereign truth that children need to be raised by both of their biological parents who themselves dwell together in some degree of mutual respect. If you can count the stars you can count the anecdotes in human literature and our own lives that support this obvious conclusion. Children that are not raised by both of their parents, regardless of the circumstances, are injured to one degree or the other. Yet in the best of times and healthiest of cultures fathers and mother die young, marriages fail, and few marriages are perfect. But by openly encouraging women to put their own happiness first and championing divorce *Feminism* has exploded the number of broken homes, and these broken homes have produced a tidal wave of injured, semi-dysfunctional, and broken people. Likewise, *Sexual Liberation*. Sexual passion is one of the fundamental characteristics of our kind and has always and will forever impact our choices and be a primary fount of both fruitfulness and contention. But Feminism's *free sex* bargain with man in exchange for equality in every sphere, and *Sexual Liberation's* destruction since the 1960s of common sense restrictions on this most powerful of impulses, has produced in the wake of our divorce culture the second great cultural marker of this age. Namely, hordes of illegitimate and often unwanted children (who slipped through the abortion dragnet). Like children from broken homes, unwanted children are a new American type—damaged kids that have had to be rescued by aunts, uncles, grandparents, adoptive parents, foster parents, and the US taxpayer. Some are repaired. All too many are not.

Sexual Liberation and Chastity

Sexual self restraint is critical for the stability and security of any *people*. Chastity is a sovereign virtue. Its fruit is simply good. *Sexual Liberation's* war on it may have produced some pleasure, but its primary fruit has been far more emotionally bent, warped, or broken men and women than there otherwise would have been. Due to sexual excess, more Americans are incapable of forming lasting marriages or finding contentment within marriage because they have made themselves incapable of satisfying or being satisfied. Our openly promiscuous culture has been especially damaging to girls because of their uniquely feminine nature in which chastity is a natural fount of honor and self-esteem. *Sexual Liberation* hasn't led to a more fulfilled American woman or made her equal with man. It has, however, made the self-loathing American woman a stereotype.

Man has fared no better for it is woman's inclination to chastity born of her *healthy-nature* that in a healthy society acts as a natural break on his inclination, born of his *unhealthy-nature*, to copulate like a feral hog.

Man and woman were not designed (or suited by *Accident* if you prefer) to be promiscuous. It is very harmful for both of them and not just because of STDs.

Abortion

In order to sustain *Sexual Liberation* as a practical reality, *Feminism* has encouraged women to do that which is most contrary to their nature, i.e., to destroy their own offspring. How much anecdotal evidence do we need to finally believe that abortion is profoundly harmful to women and that its occurrence on such a massive scale has not poisoned our culture?

Masculinity and Fatherhood

Masculine, loving, and present fathers are needed in order to produce a fruitful family. To remove them is to remove the beef from the beef stew. *Feminism's* war on the father as the head of the family and on masculinity in particular has resulted in a war on little boys that has produced generations of confused, ignorant, weak, effeminate, frustrated, discombobulated, or unfulfilled American men. These men are a fount of dysfunction who struggle and often fail to be the fathers their children need or good husbands to their equally dysfunctional wives.

This war on little boys has also all too often produced hateful, violent, or unhinged men. Our prisons are full of them.

Femininity, Wives, Mothers, Home

Woman's own *healthy-nature* inclines her, and the natural world demands, that she put the needs of her husband and children first, just as man's *healthy-nature* inclines him, and nature requires him, to put the needs of his wife and children first. Sacrifice is at the heart of human survival and progress, and the mutual self-sacrifice of the husband and wife is its first and most fundamental form. This self-sacrificing wife and mother is a principle fount of human good, the foundation of man's progress, and the central pillar of his works.

Feminism rejects this. Instead via propaganda and miseducation it has poisoned generations of American girls, making armies of masculinized, self-centered, or bitter women unwilling to submit to a husband or fully enjoy the home that they would make. Women unwilling to endure flawed husbands or forgive their mistakes, or suffer through the rough patches inherent in marriage or the midlife crises common to both men and women. It has caused women to despise making a home, the principal product of a woman's life and the greenhouse and shelter of us all. *Feminism* has despised the femininity that man so craves and needs, and has ridiculed that feminine sweetness and gentleness which are the wings under which a woman's children grow up to be strong, emotionally stable, and useful to their *people.* This woman has been the exception in film for two generations. The masculinized, sexualized, or charmless has been ubiquitous.

Drugs and Pornography

All ages produce problems that must be contended with. Chemical abuse and fornication are two such timeless problems, since man would rather not go through life sober *all* the time and will get his jollies where and when he can. However, the *Scientific Revolution*, the *Industrial Revolution* and the globalization of the marketplace have radically increased the potency and variety of narcotics. Via photography the marketplace has radically increased the availability of pornography and, since the advent of the Internet, everything that man or woman can imagine that might awaken dormant desire, from the titillating to the depraved, is now available to everyone, and is more than ever tempting and inflaming the passions.

There is general agreement in this country among *Conservatives*, *Progressives*, and users alike that the abuse of narcotics is the source of all manner of social ills and health problems. There is not so much agreement regarding pornography, though it is obvious enough that unrestrained sexual activity is emotionally as well as physically unhealthy. However, in the case of internet pornography there is now an ocean of anecdotal evidence that cries to heaven that it can stunt, warp, and bend sexual desire and seriously impede one's ability to be sexually satisfied within the bonds of marriage.

The social ills caused by these two problems today are manifest, but they have been made vastly worse than they otherwise would have been because rootless, dysfunctional, and broken men and women are far more susceptible to the temptation of easy pleasures than are the rooted and emotionally sound. *Feminism* must bear the lion's share of the blame for this because it is the single greatest fount of dysfunctional men and women in America today because of its war on human nature and thus the *natural-rational-family*.

The dysfunction produced as a result of *Feminism's* war on human nature has made the contemporary problems of narcotic abuse and pornography vastly worst than they would have otherwise been.

The Welfare State

Self-sufficiency is good and necessary as is the lending of mutual aide. A healthy *people* do indeed take care of their own that are in need. But *Feminism* has been a chief supporter of the modern welfare state not so it could help the down and out, but rather that it might liberate woman from her material dependence upon a husband. The intended consequence has been fewer life-time marriages. The unintended consequence has been the cultivation of mass sloth and corruption, and the collapse of the work ethic of the uneducated, unskilled laborer, who robbed of the virtues born of work has become morally impoverished and thus incapable of rising.

Feminism's support of the modern welfare state has helped produce a new fruitless American slave.

♦ ♦ ♦

The fruit of Feminism has been confusion, dysfunction, and misery on a an epic scale that can hardly be exaggerated. We see it all around us everyday as well as in the mirror. In its determination to eliminate the distinctions between man and woman, and to liberate them from their dependence upon each other, *Feminism* has made naturally occurring and inevitable problems vastly worse than they would have been otherwise.

Feminism has pounded the square peg into the round hole, damaged both, and declared victory. Not satisfied with having wrecked the American family and thus American culture with its *bad ideas*, *Feminism* is now wrecking the United States military.

PART THREE:
WHY WE SHOULD NOT INSTITUTIONALIZE WOMEN IN COMBAT

IX. THE INEVITABLE
REALITY OF *PEOPLES*

IMAGINE YOU ARE IN A GREEN bottom split by a winding creek, surrounded on every side by low hills. The sky is a cloudless blue. The landscape is broken by tree lines, fences, and small oak-planked, shingled-out buildings you would expect to see on old hardscrabble Southern farms. You might feel the warm breeze on your face and hear the chirping of birds except for the pandemonium of ten thousand people running for their lives. The creek is choked with corpses and runs red with blood. From what are they running? From whatever scares you most: werewolves, zombies, Mongols—a horde of club-wielding, sentient, bipedal feral hogs in leathern jerkins? Death reigns, and cries for mercy are everywhere unheeded.

Splattered in porcine blood and striking a heroic pose, you survey the bedlam while choking the squeal from your last conquest that still hangs kicking at your side. What to do?

On the surrounding hills you see growing clusters of people who appear to be successfully uniting for their common defense against the bloodthirsty feral foe. Tough as you are and already a victor in half a dozen single combats, in this swirling, squealing melée, you know that if you're not going to swell the bellies of these undiscerning omnivores you must somehow reach the relative safety of one of these hilltops to take shelter within its forming phalanx and lend what aid you can, which might be substantial and possibly immortalized one day in song.

But which hill do you choose? Fortunately, on each hilltop in the midst of the growing crowds of people waves a flag that will help you decide.

On one hill you see Taiwan's white-on-blue radiating sun on red field. From the furious melée atop the hill, the aroma of *lu rou fan* competes successfully with the foul stench of the besieging swine army, and through the shrill roar of squeals and kihaps can be heard "Kung Fu Fighting" blaring from monstrous speakers set up atop a bamboo tower hastily erected for the purpose.

On another crowded hilltop you see the red and green of Belarus, around which arcs a shower of flaming molotov cocktails made from 750ml bottles of Smirnoff hurled by tear-stained men bewailing the sacrifice, gasoline not being available and the blood of their drunken dead not being quite combustible enough.

On others you see the black and white stripes on blue field of Botswana and the May Sun of Uruguay.

The great bear of California waves defiantly behind a palisade of surfboards, the earthy smell of marijuana beckoning you.

Lastly, behind a barricade of hay bales, the blood-soaked Confederate flag waves in a breeze laden with immodest hootin' and hollerin' and the mingled aromas of the pig sty, bourbon, and barbequed feral hog.

Never mind who you are or where you are from. It is the hour of judgment. You are faced with life or death. You must either join one of the hilltop companies or be reduced to Hog Chow. How you decide what hill to make your stand on is not so mysterious: you will rally with the *people* with whom you feel the most affinity, or whom you fear the least.

That these groups formed as they did is the result of human nature. Humans are social beings, and in times of danger will naturally coalesce, starting with their immediate family and expanding outwards, even to the point of grand alliances among nations.

In this situation the people on the hilltops will tend to their own first. If prudence suggests or necessity demands, they will seek help from other hills. If their own help is sought, they will offer aid if it is to their advantage or if spurred by some cultural inclination.

The fact that *peoples* exist is an inescapable reality resulting from a stingy and unforgiving natural world that forces us at all times to strive to tame it in order to survive (the first imperative). Our nature as humans inclines us to unite with a member of the opposite sex (the second imperative), thus creating that most basic of human unions, *marriage*. Thus united its two members support one another in their struggle to survive, and also produce the offspring (the third imperative) that will perpetuate their own selves into the future. While we place the most value on this *natural-rational-family*, which is the product of this union, the danger and scarcity inherent in the natural world compels us to unite or remain united with our extended family and with those around us for greater security. Over time and continuous intermarriage, commonalities increase until you can say, "Look, there, between those two rivers! They are a *people.*"

And thus new *peoples* have been formed since the beginning of time and will never cease to be formed for as long as people live anywhere.

This reality is the inescapable fount of all that unites and divides us as humans. Why one *people* eat fried chicken and their neighbors enchiladas. Why one *people* speak French and their neighbors across the river speak German. That which creates unity among a *people* also gives rise to differences between one *people* and another.

This is the nature of things. But *Progressives* hate it because they see the differences and competition amongst *peoples* as one of the great sources of pain, and according to their doctrine, pain is wrong or evil. If only there could be one *people,* one *United Nation,* or even *one world order* then the pain brought on by the conflict of *peoples* would cease. Sounds good! Except that it is a dream within a fairy tale.

Self-interest, desire, and envy will always be hallmarks of man. Time and geography will forever divide man into distinct self-interested *peoples.* Survival will always be uncertain. The unexpected will always happen, man will always look to his own kind first, and *peoples* will always rally under leaders and ideas in order to rob other *peoples* or keep themselves from being robbed. From marauders and tribesmen to grand international coalitions, where there is profit to be had, man will rally his own to take it and face men who have rallied their own to defend it.

Conflict upon this planet is inescapable, whether it be between convenience stores facing off at a busy intersection, universities fighting for grants, or armies in a death struggle. To compete is both in our nature and required by nature.

Any given *people* or culture can be disinclined to aggression and embrace high ideals in its dealings, and this is good. But that same *people* must at all times be prepared to defend against liars, cheats, thieves, cutthroats, and feral hogs, be they lurking in the shadows or at the head of an invading army.

A *people* that cannot or will not defend itself will be subject to the will of another *people,* and this can be disastrous.

X. THE UNALTERABLE
REALITY OF WAR AND COMBAT

A STRONG MAN, known in his parts as JD, stands upon a bluff on the edge of the Rock Country looking across a river at a great fertile plain covered in well-tended pastures, orchards and vineyards. It is a prosperous land that he covets, a desire made all the worse by his conviction that the land ought to have been his, and by the memory of youthful insults and unrequited love. He hits upon an idea.

He calls the men of the Rock Country together and, being a man of great standing, there are few who do not heed his summons. Standing above the assembly upon an old stump he makes the following speech:

"Hey fellers! For too long we have suffered the arrogance of the Flatlanders and watched them grow fat on land that by rights ought to have been ours. We are stronger than they. Let us cross the river to claim what is ours, slay those who oppose us, force those that do not to tend our pastures, orchards and vineyards, and otherwise enjoy the fair fruit of righteous conquest!"

The Rock Country men immediately see the justice and benefits of such a plan and in one voice declare, "JD, we know and trust you! Command us, and we will obey!"

With JD at their head and with the full endorsement of their barefoot women who shower them with pine needles as is their custom, the Rock Country men set out down the winding trail to take what by all rights ought to have been theirs.

Some time later across the river, with bells pealing throughout the Flat Country, a breathless messenger finds a strong man known as Slim returning in haste from his orchard where he had been gathering peaches. "JD and the Rock Country men," declares the messenger, "are crossing over to kill us all!" Slim rushes to town and straight to Court Square where all the Flat Country men are gathering under arms. When he finally ascends the Courthouse steps, the throng hushes itself in acknowledgement that Slim ought to be the one to rise and speak, for Slim is a veteran of many battles and is considered their greatest man.

"Hey fellers! JD and his Rock Country men are intent upon crossing the river to claim as their own our pastures, orchards and vineyards, to slay us who stand in their way, to put to work the cowards that don't, and to make serving wenches of our women. They are as fierce and wild as the raccoons they fill their bellies with, but we have right on our side and are better men. Who will away with me to the river to drown them in it?"

The Flat Country men immediately see the justice and necessity of such a plan and in one voice declare, "Slim, we know and trust you. Command us, and we will obey!" Swollen by the encouragements and cheers of their women arrayed in sundresses who shower them with magnolia petals as is their custom, they set off for the river to find JD and the Rock Country men disembarking from their bass, pontoon, and flat-bottom boats.

The Battle of Pebble Landing is fierce and cruel, as battles usually are when *Rage*, *Panic*, and *Chaos* are present. The face of the enemy and presence of so many friends, along with full bellies and the absence of wounds, fill both sides with hope and courage. The Flat Country men are driven by desperation to keep safe their families and to keep possession of their fathers' labors. The Rock Country men are driven by their hunger for land, by the river that now hems them in, and by a thirst for the fruits of victory.

In a rush of howls, JD's and Slim's *people* now close with each other and soon have appointed *Chaos* the general of both. *Rage* and *Panic* fly to and fro vying for lordship. The struggle to kill and not be killed squeezes dry like a sponge every ounce of strength from every man. All is poured out. Nothing is left in reserve. The stakes are too high.

The costs for both sides are great. With each fallen man a thread in the fabric of society is snapped, leaving children unconceived and widows adrift, and only memories to guide and protect fatherless children. The fallen are dead minds and hands that will produce no more, improve no more, and bequeath nothing more to sons and daughters that they might be more firmly rooted.

Much will be decided in the contest: Who will tax and who will be taxed? Who will command and who will serve? Who will strut and who will doff the cap and seethe in hate and shame?

This in a nutshell is the nature of war and combat, and it will remain so forever as long as envy, vanity, greed, lust, hate, love, self-interest, shame, fear, belief, and want are inborn human traits.

History and common sense teach us that in the affairs of *peoples*, circumstances will from time to time raise up the man who possesses a vision that would put his *people* at war with another, and that other men will follow him. History and common sense also teach that the nature of war is often one of extreme brutality and that predicting its particulars and outcome is difficult. History likewise teaches us that war can produce consequences for a *people* that range from total victory to apocalyptic defeat.

It is an unbreakable rule of nature that the strong survive and the weak survive by leave of the strong. If *We the People* are to live on our terms then, be we the aggressor or defender, we must at all times be prepared to present our strongest possible side to the wars and enemies that lurk in the unknowable future.

And in the realm of human combat where the blade cuts skin, that side will always be the exclusive realm of young, strong, martial men. And there are unalterable reasons for it.

XI. WHY COMBAT
BELONGS TO MAN, THE AGGRESSOR

ALLSTATE INSURANCE RUNS a commercial where a man and a woman are sitting together in some café and the woman challenges the tongue-tied man's assertion that men are superior drivers, producing her Allstate Safe Driving Bonus Check to prove the point. This is classic obfuscation and misinterpretation of the obvious. Collectively speaking man is indeed the superior driver. Man has not only built virtually everything that stands and moves, he is also the tool-maker of our kind and wields them best. He cannot boast of, and she ought not be ashamed or envious of what God (or *Accident* if you prefer) settled long ago.

The evidence is ubiquitous. Look behind the wheel, the helm, or the steering thingy of most every complicated piece of machinery, and in the overwhelming number of cases you will see a man. Where the operation is complicated or dangerous, there he'll be. NASCAR is dominated by men for this reason (and others) and it is the rare woman who can compete in that arena, and exceptions do not disprove the rule.

What the ad could have asserted more truthfully is what is implied within the title Allstate *Safe* Driving Bonus Check i.e. that women are *safer* drivers. Broadly speaking this may be debatable but in the case of young people it is clearly true. Ask any actuary.

Show me an unmarried man much under twenty-five behind the wheel and I'll show you a spontaneous-misdirectional-road-torpedo. Young men are forever getting themselves killed on and off-road, but it's not just behind the wheel that they represent such a threat to themselves and others. Generally and relatively speaking, young unmarried men are just far and away humanity's most aggressive and volatile demographic.

They are so aggressive and volatile because man possesses an inborn and uniquely masculine spirit that tempts, urges, compels, or drives him to risky behavior.

It is this uniquely masculine spirit that can easily be seen in little boys on the playground and can certainly be seen in groups of young unmarried men. Unrestrained, this spirit can manifest itself by acts ranging from innocent horseplay to full scale riot. In a matriarchal society this inclination towards aggressive behavior in young men is a guarantee of disorder and insecurity, as we see in our inner cities today.

But in a society where this impulse that compels young men to disorder or risk is restrained by older men, this masculine assertiveness becomes a positive good, and a primary source and driving force of the *people's* progress and security.

In the natural order of things it is first the father who begins the process of taming the spirit of his son and shaping him into a useful man acceptable to his *people*. He is supported by the men of his tribe, including older boys, coaches, teachers, preachers, and sometimes even law enforcement.

With his youthful, impulsive spirit thus tamed, the young man can join the men of his tribe when they sally forth to hunt the wooly mammoth. Or he can play his part on the line of scrimmage before his *people* on a Friday night. He can also succeed in his studies, that most unnatural of modern preoccupations. He will try and succeed at the hard thing rather than fail due to distraction, boredom, or sloth. He shows up to work on time and is valued by his employer.

He helps the old lady across the street instead of robbing her.

And down South when he is spoken of he will be called that which every father would have said of his son: *He's a good boy.*

But more to the point of this treatise, the high-spirited young man that has been tamed and pre-hardened by the men of his tribe is the new patriot and ideal material that drill instructors in our armed forces can hammer into disciplined fighting soldiers, sailors, marines, and airmen.

In the military setting, he can be improved and hardened by strict and even fierce discipline at the hands of his *Captain* who will curse, insult, and beat him as necessary in order to drive him to do the thing he does not want to do, fears doing, or does not believe he can do. But when he succeeds, his *Captain* will slap him on the back and say, "Pillsbury! You may be a man after all," at which point the pain is forgotten and replaced by strength, confidence, and self-respect. To turn boys into men you break them down and build them up, over and over. Like folding steel.

This masculine spirit is at the very heart of combat, and it should go without saying that only men possess it. And exceptions don't disprove the obvious truth.

But man possesses another uniquely masculine characteristic that is an extension of his masculine spirit and, though normally latent, is the very black, bloody heart of combat itself. We are speaking of battle-rage or battle-madness or *going berserk*. It is the opposite of panic, a state of mindless fury that is only stopped with blunt force or a lull that allows the mind to perceive once more. It puts the din in battle. Women and weak men cannot stand against it. A strong man stands against it with caution. Ask any cop.

This madness has been the deciding factor in battles and engagements beyond reckoning, and successful *Captains* since the dawn of combat have carefully cultivated it within their men that they might be able to unleash it at the decisive moment, often by means of fiery words and speeches. Picture a football coach firing his team up before a game, or the king crying out to his men, *Once more unto the breach!* Or the most oft repeated and shortest combat speech of all time in every language: *Charge!*

Like male bonding, this masculine rage-potential is real, it's a man-thing, and a key element of combat, as critical on the battlefield as supply lines and sound tactics because sometimes, when the *Captain's* plans lie shattered and *Chaos* runs unleashed, it is the fury of his men that saves the day.

But in addition to man's natural aggressiveness and the rage-potential born of it, man possesses a unique and acute anxiety that further makes battle *of man* and forever his domain.

Consider *Gallipoli*, a 1981 film named after the peninsula of Gallipoli, which is west of the Dardanelles and part of Turkey's European side.

During World War I between April 1915 and January 1916, Great Britain and its allies, including Australia, fought and lost the Battle of Gallipoli against the crumbling Ottoman Empire, an ally of the Central Powers. In the movie we follow two young Western Australians who, by the end of the film, find themselves with their regiment in a trench preparing to "go over the top" to assault the Turkish occupied trenches across no man's land.

The first wave is cut to pieces by rifle and machine-gun fire, as is the second. Due to tragic misunderstandings and miscommunications the regiment must continue the fruitless assault. In a pitiful scene we see an Australian soldier hook his wedding ring onto his bayonet that he has driven into the trench wall. He knows he is about to die, so he leaves this token of his life hanging on the bayonet that he will not get close enough to the Turks to use. With trembling hand the regimental commander raises his whistle to his lips to give the signal to charge. This time he'll go and die with his men. And die they do.

What do we imagine that the mothers, wives, and daughters of all the hundreds of thousands of men killed after climbing out of the trenches during that terrible war would have said had they been able to speak to their man in the final seconds of his life? Though they might speak with a bit more hysteria than I'll attempt to convey, I imagine them crying out, "Don't go! Stay down!"

But why did they go over the top? The reason that a man will stand and face danger or even march to near certain death is because he possesses a uniquely masculine anxiety that woman does not.

He fears being thought a coward.

Men understand instinctively that they must have their friends' backs. To fail to answer their call for aid, to flee while they stand is to exile oneself from one's own tribe, to live alone in the wilderness, despised by those you love as much as you despise yourself. For a man to be thought a coward is mental castration. It is living death. It is the IV drip that poisons all of his works for the rest of his life unless it be washed away by conspicuous valor.

This masculine fear of being thought a coward is one of the primary founts of courage that has helped keep terrified soldiers on their feet and facing the enemy for ages, and will for ages to come.

But this fear of being thought yellow by one's friends is matched if not surpassed by the fear of being thought so by one's own woman.

Imagine a man, woman, and their child sitting together when suddenly a great danger appears. Now picture a charging 325 pound squealing feral hog, all muscled-up and angry, brandishing a two-sided axe. It is the rare mother that will not gather up her child and flee, expecting her man to protect them. She can abandon him, screaming with tears running down her cheeks and no one, anywhere, in any age would would think her conduct shameful. But if *he* abandons *her*, he would suffer universal condemnation by all people, in all times, by both man and woman alike.

He is a coward.

Just as woman has been granted tears and sobbing to assuage her inner turmoils, God (or *Accident* if you prefer) has given her a flight instinct in the face of danger. It is this instinct to flee danger as well as to tend to the immediate needs of her young that has made it possible for the daughters of Eve to keep the race alive in the world into which we were cast—or sprung up if you prefer.

But not so man. His instinct born of his *healthy-nature* is to stand and defend his woman and children, and when *Panic* would grip his heart and tempt his *unhealthy-nature* to flee, his fear of being thought a coward in her eyes, an instinct equally born of his *healthy-nature*, will keep his feet planted towards the enemy.

Because to be thought a coward by her—well, Viagra can't fix that.

Combat is what it is because it is a manifestation of man's unique masculine nature. It is what it is because he is what he is. It is this nature that enables him to prevail or survive on the battlefield. It is this nature that compels him to live dangerously or enables him to face or attack danger. It is his inborn fear of being thought a coward by his friends and woman that enables him to face or attack that which terrifies him.

Woman cannot stand up to combat amongst men because she is not man. She is woman and possesses her own nature. Hers is to retreat from danger, his to approach it. Together they survive.

This is Human Nature 101 and the proponents of institutionalizing women in combat simply ignore it.

XII. THE *CAPTAIN*
AND HIS BAND OF BROTHERS

IT'S A SUNDAY MORNING and you, along with so many others in your Southern county, have elected to go to church, and there you sit listening to a sermon about being good and not being bad when suddenly the front double doors burst open and in rushes a breathless man who screams. "Godzilla! It's Godzilla!"

Outside you hear a deafening roar, the crashing of trees, and the splintering of trusses. When you finally make it to the door, sure enough, there is Godzilla with his head stuffed in the now roofless Dairy Queen across the street, eating employees and patrons alike. Yes, they should have been in church.

Up and down the street all pandemonium has broken loose, and the wail of tornado sirens on such a clear sunny day only adds to the confusion and leaves folks in the countryside scratching their heads. In fairness to the authorities they weren't expecting Godzilla.

Ever the drama queen, Godzilla raises up to his full height, rears back his oversized head and makes as if to give a great triumphant roar, but is embarrassed when an Abbelch four octaves below the staff escapes, heard and smelt a mile away.

Now if something is not done to stop Godzilla, he is going to flatten all the churches and restaurants in town, and when he can't eat another bite, he's going to stomp on people the way a child stomps on bugs—just for fun.

So what's to be done?

It is possible that some stout-hearted woman will flip a great oak table, break off a leg, run up Godzilla's back, and beat him with it. It's just not likely.

But if the people do survive, here is what will happen according to the script written on our DNA by God (or *Accident* if you prefer) eons ago:

Instinctively, the women will gather up their children and without shame flee and hide if they can, and only fight if they must. The children expect this and their mothers will not be parted from them. A woman's flight instinct is what keeps the race alive.

The men will first and foremost see to the immediate safety of their own women and children and attempt to put them out of harm's way, and their women expect this. A man's instinct to protect his woman and children is what keeps the race alive.

Before and after he has secured his woman and children a man will instinctively look about him for other men to ally himself with, and he will do so with any man his instincts tell him can be trusted.

At the same time something almost magical will happen. One man will rise up and start giving orders, and all the other men will obey him. He'll usually be older, but as a rule not an old man, but whether he is thirty or sixty he'll be an active man that exudes confidence and appears to have a plan. He'll be that alpha male, half born and half made, that all *peoples* produce. Men will follow him because when in danger men are happy to be pawns as long as they believe their king can lead them to victory. Thus led, working together they will secure their families, stand guard over them, and perhaps even sally forth as one to confront Godzilla.

There is no guarantee that a competent leader will emerge or that the men will organize an effective resistance, but if they do not everyone is Godzilla Chow.

The survival of all of us is at all times dependent upon this paradigm of armed men under the command of a *Captain*, be that *Captain* a tribal chief, king, governor, or county sheriff. In times of peace this paradigm exists 24/7/365. In our country today all you have to do is dial 911 at any hour and scream "Help!" Where this paradigm does not prevail there is no peace and disorder and insecurity are the best that can be hoped for. In times of war or when Godzilla attacks, defeat and death are inevitable.

Just as ants have a nature, man is hard-wired to work and fight in groups of men, and to give or take orders, whether on a construction crew or in a 500,000-man army invading Russia. Likewise woman is hard-wired to look inward to the needs of her dependents, whether they are her children, the sick, the elderly, or her fighting man himself. There is security within the family because she is present. Likewise there is security within one's *people* because he is united with the other men of his *people* under the leadership of a single man. Determining who this leader will be is not always done peacefully, but men can quickly establish pecking orders and can mend fences just as fast.

Within the groups of these working or fighting men bonds of trust and sentiment unique to man are formed. This is male bonding. Within the odiferous radius of Godzilla, lifelong bonds of friendship can be welded in a moment. This is male bonding in its purest form. It is the band around the *band of brothers*. It is real and critical if Godzilla is to be defeated, and the butchest woman can't do it. It is indeed a man thing. Male bonding occurs naturally and must occur if the group is to perform its function well. The more difficult or important the work, the tighter the bond must be. Where the bond is weak or no bond forms, the work suffers, be it herding cattle on the prairie, framing on a construction site, or a barroom brawl.

This male bonding is a bone-deep human reality and only fools mock it. It is not a social construction, and in combat, where the stakes are life and death, it is of supreme importance. It is not only critical to success in battle, it is part of the bedrock of morale generally. Technology, strategy, communications, tactics, and supply are all necessary for success in war, but morale is first among these equals.

♦ ♦ ♦

The family is the fundamental building block of humanity from which *peoples* naturally emerge. These *peoples* by nature will inevitably struggle against each other. Combat is the most extreme, violent, and consequential form of this struggle and it is born of man's unique masculine nature.

For these reasons American combat units must remain composed of young, strong, martial men and what older, active, alpha males are required to lead them. As they have always been throughout all of human history.

To add children, the lame, the sick, the elderly, the cowardly, or women to the ranks of our combat units is to mix clay with iron, and we should assume that our future enemies will not be so foolish as to include them in their armies' ranks, but if they do we'll thank our lucky stars as we take them to the woodshed.

XIII. WOMAN'S NATURE
AND THE NATURE OF COMBAT
ARE IRRECONCILABLE

THAT WE EVEN HAVE TO POINT out the nature of woman that makes her so ill-equipped to endure and succeed in the human arena called combat is a sad sign of how deeply one of the two basic premises of *Feminism* has sunk into our collective consciousness, including that of so called *Conservatives*.

Man and woman are equal. Period.

And the round peg will be driven into the square hole.

Women's suffrage, improving woman's legal status, and expanding her work opportunities is one thing, but women in combat? It's akin to gay marriage. It leaves us scratching our heads, asking ourselves how in the world we slept through the chain of events that brought us to this point. When was nature rewired and reason inverted?

No, man and woman are manifestly unequal, but neither is one superior to the other. They have distinct yet complementary natures that when joined in marriage create the basic building block of human society—the *natural-rational-family*.

Just as home is what it is because it is an extension of woman's nature, combat is what it is because it is an extension of man's nature. He does not need to alter his nature to enter this arena. She must.

It is understandable that opponents of women in combat focus on the obvious physical weakness of women relative to men, but by over-emphasizing this argument we have allowed proponents to play their only face card, off-Jack though it is. The unprecedented mechanization of war and the United States' absolute (though temporary) military supremacy have indeed made combat for U.S. soldiers *right now* less dangerous and physically demanding than it was during—let's say—the Battle of Kursk. But all this moment in time has done is to allow *Feminism* and its *Progressive* coreligionists to create a dangerous illusion. Physical strength and endurance will forever remain a decisive factor in combat, but as obvious as woman's physical limitations are, it is her mental and emotional nature and the impact she has upon man that has been and continues to profoundly weaken the fighting culture of our military.

First, please, let us agree that we must have men and that no one is seriously pushing for all-female combat units. Assuming agreement on this point, let's now consider the reasons why woman's uniquely feminine nature makes integrating women into combat units foolish.

Distraction and Male Bonding

I posit that the US Navy does not allow its submariners to smoke marijuana on-board their expensive boats. I am so confident of this that I am not even going to do a Web search to verify.

And why not? Obviously because it would affect their state of mind. Not that a few tokes would render them useless—but certainly a little less alert and clear-headed than you want sailors to be who are floating around underwater in one of the most expensive machines ever built with a belly full of nuclear missiles.

The *Captain* is not going to allow on board anything that could seriously distract his sailors or impede their ability to perform their jobs, and there is nothing more distracting to man than woman (except maybe pot to potheads).

That woman influences the behavior of man (and vice-versa) shouldn't have to be pointed out. The interaction of man and woman is the heart of romance, but it is a heart that—besides butterflies, distraction, and love—also produces anxiety, resentment, jealousy, contention, division, and sometimes tragedy.

Wherever the paths of man and woman cross, the possibility exists that he (or she) will be distracted from the task at hand. From the classroom to the factory floor to the office, blossoming love affairs both proper and improper are legion. This must be contended with in our day-to-day lives, but there are times when this distraction must not be permitted, namely prior to and during an event that requires a high level of focus and teamwork among men. A football team preparing to play would be a good example, but combat is the best because of the stakes involved.

Male bonding is real and absolutely essential for men who would work together. It occurs when men are together without the company of women, whether playing, working, just visiting, or fighting. It occurs as they come to trust each other or learn to work together. Women can believe this is true because they can see its equal in female bonding, which occurs in groups of all women, and like male bonding, it is born of her DNA and is not a social construction. It is a uniquely feminine and needed source of pleasure and sympathy that can only be found in the company of other women who can understand and connect with one another at levels that man cannot. Besides the common work that such bonding can advance, it reconstitutes them all to face the work and challenges of their lives.

What female bonding does not prepare them for, however, is a bayonet charge.

Women know that the introduction of just one man will change the dynamic of a gathering of girlfriends; likewise, the introduction of just one woman will change the dynamic of a group of men. If the intruder is sexually desirable this can be especially disruptive as both man and woman are by nature highly sexually competitive. Man is very much like his counterparts in the animal kingdom in this regard.

Men need the company of their own sex, especially when there is hard and complex work to be done. Neither can man escape his sexual inclinations any more than woman. By excluding woman from his combat unit, the *Captain* avoids a needless distraction and protects his unit's cohesion.

Woman is indeed highly distracting and a solvent to male bonding.

The Cost of Accommodating Her

War has always been extremely expensive. It takes productive labor out of the economy that must then be supplied and can be highly destructive in terms of lives and property.

Whether he is raising an army to defend his realm or to conquer the one next door, the king must count the costs. He'll usually not be able to afford all that he needs and he'll never be able to afford all that he wants.

He'll spend his coin on what he absolutely must have and make do elsewhere.

What he is not going to have is unnecessary expenses.

Considering the tiny number of women that could possibly make effective combat soldiers, the added expense of accommodating the fairer sex is coin very poorly spent. Woman's medical needs alone make it prohibitive and the king's *Captain* in the field should not have to be the least bit concerned with the myriad of medical issues unique to her.

Then there's the issue of modesty. The *Captain* should not have the added expense of providing separate places to bathe, dress, relieve oneself, and sleep. Unless of course he is just going to throw his female soldiers in with all the other men, which is exactly what he'll have to do when the bullets start flying. But we'll talk about her degradation later.

The Costs of Protecting Her

Securing the well being of one's men is half the nature of war. After all, war can be boiled down to two primary components: attack and defend. The Captain must dedicate resources to food, clothing, shelter, medical care, and protective gear. These are all unavoidable expenses.

Likewise, he must also take measures to minimize injuries. The combat theater is by its nature a dangerous environment even when the bullets are not flying, as mental strain, physical exertion, and the nature of war's tools and material will converge to produce a higher rate of injuries than the most dangerous of peacetime occupations.

A woman in this environment is at much higher risk of injury (unless she is given special consideration) because she'll be subject to the same blows, bumps, and scrapes as the men, and a tough young man can shrug off blows that could seriously injure a tough woman. High school football comes to mind, which incidentally must be more dangerous than combat, for who in his right mind would put a woman on the line of scrimmage against sixteen year old boys?

The *Captain* should not be asked to expend resources to make his combat theater as safe for women as it is for his vastly stronger and tougher men, or risk the loss of a female soldier to injuries that a man could more easily endure.

The *Captain* will also have to expend resources on a police force to bring to heel criminal elements among his men and to maintain discipline. This too is an unavoidable expense. But what is avoidable is the expense of having to protect a handful of female soldiers from being raped by his own men in the semi-lawless environment of the army in the field.

Imagine a man going into a typical redneck beer joint on a Friday night where he proceeds to comment loudly on the fine female rear ends he sees present. His attention being drawn to the bar, he strides up where sits a redneck who has a lovely specimen leaning on his arm and demands to trade places with him. When the astounded redneck rises, faces this intruder upon his besotted revelry and says, "What the #@%^!," the intruder promptly takes the redneck's red plastic cup of "Keith Stone" Light and throws it in his face.

Such a display would certainly unite the other rednecks in opposition, and if the intruder is extremely lucky he'll just find himself escorted to the door. But more likely a quick and decisive brawl would ensue that might even be followed by police lights in the parking lot and would certainly be much discussed for weeks on end.

The truth is that this doesn't happen very often because fools are not long for this world. Boys learn early on the playground that you just don't act this way.

Now imagine a young woman getting all done up to go out on a Friday night. When she gets to her darkened, thumping, light bespeckled club she proceeds to drink heavily and allows herself to be separated from anyone that gives a rip about her, strutting her stuff without shame. She continues to drink until she is thoroughly wasted and then on up to blackout drunk.

When daylight and consciousness returns, she discovers to her horror that she was gang raped by three drunk men, all strangers to her.

Who is to blame?

The men are, of course. They raped her, and being drunk out of their senses is no excuse for it. Their reputations and careers will be destroyed, their families disgraced and devastated, they'll be financially ruined, and will spend any number of years in prison. This outcome is necessary and will hopefully deter other drunk men from giving way to their dark impulses.

But she is also to blame.

At this point *Feminism* is blowing a gasket. How dare we blame her! She *ought* to be *Free* to get as drunk and act as outrageously as she chooses and not suffer any consequences.

The problem is that this is literally impossible. Like the obnoxious man who threw the beer in the redneck's face, we will all suffer the consequences of our foolish decisions—or for embracing the *bad idea.*

Feminism's solution is as always to fight the symptom and not the root of the problem. Through its allies in the media it will cast the blame on nightlife culture and declare that it *ought* to be made safe for drunk girls, but it might as well tell the river to flow uphill.

The solution is what women figured out eons ago. Modesty and propriety is the best deterrent against unwanted attention, and caution, circumspection, and common sense the best defense against violence. When outside the security of one's own castle, be wary of finding yourself alone with strange men, always be with friends when around drunk men, and never, ever, ever find yourself drunk and alone with strange men, be they drunk or sober.

The man who threw the beer in the face of the redneck was beaten because he was a fool; likewise, the drunk party girl.

Feminism decries rape in the military and is right to do so. Whether the victims are female soldiers or civilians in an occupied territory, swift and severe punishment should be meted out to American soldiers that rape.

But a combat theater is an inherently dangerous and semi-lawless environment under the best of circumstances. It cannot be made wholly safe for strong men and will always be considerably more dangerous for weak men and vastly more so for uniquely vulnerable women because of the natures of man, woman, sex, and rape.

The *Captain* in the combat theater should not be asked to expend resources to make his theater as secure for women as it is for his far tougher and far less vulnerable men. The solution is to exclude women as they always have been.

The Poor Return on Investment

Let's say the *Captain* goes out shopping for bombs and he has a choice between a $100 *bangpopper* that could blow a hole in a 12″ concrete wall and a $95 *zapsparkle* that could blow a hole in a 6″ concrete wall. Which bomb does he buy? As a rule he'll always buy the *bangpopper* because it costs only $5 more and has much more destructive potential.

For the exact same reason the *Captain* will always choose male soldiers over female soldiers.

The cost to train a female vs. a male soldier is for all practical purposes the same, as is the cost to transport her and maintain her in the field, despite her lighter weight. However, a man's potential value in the combat theater makes the benefit-to-cost ratio of male soldiers vs. female soldiers incomparably better.

For the *Captain*, female soldiers are a poor investment.

Her Flight Instinct

Nature has granted to woman a flight instinct that in no time or place has been considered cowardice. Her fear of danger and inclination to gather up her children and flee from it is how the race has survived since the beginning of time.

This uniquely feminine fear makes her particularly unsuited for the cauldron that is the world of men fighting. It is not that there do not exist highly courageous, masculinized women, but we are talking about *institutionalizing* women in combat. There are simply not enough of them to make the vetting process anything but highly impractical, inefficient, and expensive.

Feminism will of course express outrage at any suggestion that woman is not as courageous in the face of physical danger as man, yet it spends a great deal of time in its cultural war trying to convince wife-beaters to quit being bullies and women to stand up to them or at least seek out others to intervene on her behalf. If woman has for umpteen thousands of years cowered beneath the threats and blows of her caveman husband, why do we suppose that she is suddenly going to stand in the face of wild-eyed twenty-something men sanctioned by their government to kill her?

She is Averse to Risk and a Short-Term Thinker

I don't know how many gaskets *Feminism* has, but I am going to go ahead and blow them all.

The overwhelming number of strategic thinkers have always been, are now, and shall forever be men. Likewise discoverers, explorers and inventors. Likewise monarchs, presidents, governors and CEOs. Man should not take pride in this and neither should woman be ashamed or resentful. It is just the nature of things.

This is not to say most men are movers and shakers, they're not. Or that there are not some women that are. But the reality is that man is more likely to take initiative and woman by her nature is more averse to risk and more inclined to focus on the here and now.

For the *Captain* to condition his men to see the bigger picture and to risk their lives for the greater good is to train the wild stallion to pull the plow.

Woman by Nature is Semi-Independent

Given all of *history* as our source, we are hard-pressed to come up with many examples of large female-only groups working in concert, even in the modern era. But examples of all-male groups are as numerous as grains of sand on the beach. The reasons are twofold and obvious. Man is hardwired to work in groups with his fellow man to an extent that woman is not, and the reason is that just as combat is *of man,* she is hardwired for another purpose, namely to make a home.

There is no humanity without home. It is as needful to us as nests are to birds. The woman by her nature makes home. Metaphorically she *is* home. The reality of this human nest and all the needful and desirable purposes it serves requires her to do what she would do instinctively—look inward to the demands of this natural imperative. Thus by nature she lives in a state of semi-independence; she is mistress of her own domain. Man must march in step, shoulder to shoulder with his friends under the command of a *Captain* to protect the tribe. It is his nature to do so, and if he does not, the tribe perishes. Woman must look inward to the home that sustains life and nurtures the future of the tribe. If she does not, the tribe rots from within. But unlike her man who must obey his *Captain*, she performs her role far more independently and on her own terms.

Consequently, woman doesn't always play well with others. She certainly doesn't always play well with other women. That should blow *Feminism's* last gasket for sure.

The nature of combat requires a level of discipline, command, subordination, and cohesiveness that comes naturally for man working with man, but not so much for woman who by nature would rule her own nest.

He Doesn't Become Pregnant

Then there is the issue of pregnancy. Losses due to death, wounds, illness, exposure, capture, and desertion are unavoidable problems for the *Captain*, but he never loses a man, or the time and costs of training him, because he becomes pregnant. This is a difficult problem to surmount in the workforce, but in the realm of combat where the stakes are so high, and where predictability is as necessary as it is fleeting, pregnancy is not a possibility that the *Captain* should have to consider.

Woman is Emotional

Then there is the highly sensitive issue of woman's uniquely emotional nature. But the issue is only sensitive because *Feminism* has made it so. There's irony in here somewhere.

It should go without saying that man is less emotional than woman and led more by reason, while woman is led more by intuition.

Whatever this emotional nature is that has so perplexed man since Adam first asked Eve to pass the salt and she burst into tears, woman carries it with her everywhere she goes, and everyone from her husband to her girlfriends to her employer must take it into account.

Woman does not easily compartmentalize or turn off her feelings, and if she does so she runs the risk of diminishing herself as a woman, of blinding herself, for her feelings are the fuel of one of her greatest natural strengths, namely her intuition. These feelings ebb and flow faster than man's and can grow disproportionately with the intimacy or intensity of the situation in which she finds herself. In our day-to-day life this nature of hers is a fount of human drama and adds to the spice of life.

But in the combat unit, especially one in the field, this emotionalism (besides the effect it can have on unit cohesion) greatly complicates the *Captain's* work. Managing the feelings (morale) of his men is supremely important, the heart of leadership, and never simple. But at least the emotional needs of his men are simple and the *Captain*, being a man himself, understands them. All men have known since the beginning of time and all men know today that managing the emotional needs of but a single wife is delicate work, and there will be a terrible price to pay for failure. But the Captain in the field ought not have to take into consideration the feelings of wives, daughters, and sweethearts or the effect they are having on his men or on each other.

Woman Doesn't Let It Go

A 9th grade girls' basketball coach once made the following observation: "Boys can have a knock-down drag out fight and be shaking hands as friends ten minutes later. Girls wound each other with words and make enemies for life!"

Such a stereotype will infuriate *Feminists* but I guarantee that Adam and Eve, sitting on a log outside their cave, made the same observation about their great-granddaughters.

Man cannot be compared to woman for kindness, gentleness, or sweetness and such women are the foundation of civilization and the heart of all that is worth living for. They give, sustain, and improve life.

But insult her, cross her, or injure someone she loves, and there is hardly a brake on her resentment, or on her vindictiveness if she is a mean woman to begin with. Woman just cannot let it go as easily as man, and the *Captain* on the battlefield has enough drama to contend with.

Aunt Erma

And speaking of the uniquely emotional nature of woman, man doesn't have a period.

Her Best and True Nature

War paint does not become a woman, and her presence on the battlefield will evoke neither fear nor admiration in our enemies. It can only produce disdain or needless provocation.

Sweetness, kindness, and gentleness do become her and will always inspire her own men to feats of heroism, hardening their resolve not to give way.

Of our kind she is the giver of life, the healer, the nurturer, the sustainer, the adorner. She is our beautiful side, our better half. This role was assigned to her by God (or *Accident* if you prefer) and only *Enlightenment's* spawn *Feminism* would dare challenge or presume to change it, and any attempt to do so will only make her ridiculous, ugly, or prey.

◆ ◆ ◆

Woman does however possess half of humanity's nature and thus will forever play a critical role in success in battle.

XIV. WOMAN THE CIVILIZER

A WOMAN IS WALKING ALONE late at night along a dimly lit and deserted city street. Behind her she begins to hear faint footsteps and the low murmuring of a man's voice. Second by second the steps and voice draw nearer and her anxiety grows. She hastens her pace instinctively. Her anxiety turns to mild fear, so without breaking her stride, and as inconspicuously as possible, she turns her head to confirm that the man who now must be within twenty paces of her is no threat.

In the shadows she sees the bulk of a large man. Beside him walks a woman carrying a baby.

Is the woman still anxious? Not at all. Her anxiety disappears all at once, and she is at ease. But why is that? She is still walking alone on a dark, deserted city street and he is still a man.

This simple illustration shows plainly some fundamental aspects of human nature and primary reasons why putting women in combat will weaken our combat readiness.

The problem is not woman's lack of strength and endurance, though that would be reason enough. Neither is it her uniquely feminine nature that makes her emotionally and mentally ill-equipped to succeed in a realm defined by the fury and violence of men killing men.

The reasons are rather the dramatic and transforming influence that she has upon man, the powers that she gives him, and his need to be needed by her.

God (or *Accident* if you prefer) has assigned to woman the role of taming man, giving him a greater purpose beyond himself, and converting his energies to productive purposes. Where men do not have their own women to protect, and wherever undomesticated men congregate—be it a playground, frat party, a street gang, a pirate ship, or an army on the march—possibilities ranging from buffoonery to mayhem to catastrophe increase exponentially.

Upon taking a woman *to have and to hold*, the aggressive, selfish, and shortsighted instincts so associated with male youth begin to convert to the defensive, selfless, and farsighted—a transformation completed the very moment he sees his first-born child. Man possesses an inborn and natural imperative to have a woman, but once he has her, he feels a compulsion to protect and provide for her and their children in a dangerous and uncertain world.

This makes him less stupid and more reliable.

But equally he needs to be needed by her, adored by her, and served by her in turn. It is food for his soul which swells his spirit and makes him more than twice the man he would be without her.

The lower ranks of our military may be full of unmarried men, but the belief that they are fighting for mothers, sisters, sweethearts, and the adoration of their womenfolk generally will have a profound effect on their morale. Equally importantly the officers and NCOs that lead them are often married men, who as such will direct their troops' substantial energies toward the protection of the *people*.

It is mothers, sisters, daughters, sweethearts, and wives and their need of him that makes their man the *people's* best soldier.

The belief that he is protecting hearth and home, more than anything else, makes the strong man—married or unmarried—stronger and keeps the soldier on his feet in the face of pain, hunger, exhaustion, despair, and terror. Where this paradigm does not exist, there is security for neither woman nor man.

By pretending to put woman on an equal footing in our combat units we undermine this primal phenomenon and foundation of our soldiers' morale.

XV. WOMAN THE DEMORALIZER

IT'S A BRIGHT CLEAR DAY at Camp Wherewithal, where today's civilians are being turned into tomorrow's soldiers. The cursing of drill instructors mixes with the chants of marching recruits, and in the distance can be heard the cracking of the rifle range.

Our company, C Company, surrounds a great man-made pond spanned by a narrow gangway not wide enough for two men to pass without difficulty. C Company's drill instructors are running their charges through a close-combat drill which consists of two men armed with padded staves entering the gangway from opposite ends, meeting in the middle, and attempting to knock each other into the water.

The raucous spirit of the Coliseum prevails as each platoon sends its recruits across the gangway in rapid succession.

We follow a recruit named Cody, a typical male specimen of our time: a pasty-faced, pudgy, corn-syrup-fed suburbanite. Quarter-educated, a stranger to want or sacrifice, as immature as his tastes are jaded, but in fairness, the victor of a hundred thousand single combats with the most vile video-game creatures that someone else's imagination could concoct.

It is with some trepidation that Cody sees his turn to cross the gangway approach. Scanning the other side he does a quick calculation and to his horror determines that he must face Joe-Billy, a colossal 18-year-old from East Tennessee and former all-state-something-or-other. That's just unlucky!

But lo and behold if he hasn't miscalculated. The hollow-cheeked boy in front of Cody must face Joe-Billy, which he does manfully, and rather more manfully, Joe-Billy sends the boy careening into the water with a single blow devoid of compassion or restraint. Now that's just lucky.

But wait! What is the apparition that now stands at the opposite end of the gangway? Oh Lord, no! Its Megan! All 5′ 10″ and 165 pounds of South Carolina upcountry, five times Lil' Miss Judo Champion, "Axe-to-Grind" Megan.

With eyes flaming she calls to him, "You wanna piece of me?" But judging by her raised voice, intonation, and rapid forward march, Cody quickly concludes that the question was indeed rhetorical, and not an oddly timed proposition.

Hesitatingly, Cody advances to meet the she-demon, suppresses a distant echo in his mind admonishing him not to hit a girl, and takes her padded staff square in the face, not being able to dodge it if he wanted, and falls half-stunned backwards into the water.

Now, is there a difference between Cody being knocked into the water by Megan, and the hollow-cheeked boy being knocked into the water by Joe-Billy, or is each case identical, each couch potato simply being bested by a physically and martially superior opponent?

There is a difference, and it has to do with a fundamental, primordial aspect of human nature. A man can take a beating at the hands of another man and still remain a man. But to be beaten by a woman is to be humiliated, a kind of mental castration.

When the hollow-cheeked boy comes out of the water he comes out to both jeers and cheers, but he comes out a man because he faced his opponent with courage and was beaten—by a man.

Cody, on the other hand, emerges to ridicule, the butt of a joke tattooed upon his forehead, never allowed to forget that he is the boy who got whipped by a girl.

Feminism, which in its bitter heart hates man and delights in this sort of role reversal, will say, "So what! Don't we want the best fighters, be they male or female?" Such reasoning is based upon its willful and selfish misunderstanding of human nature and a rejection of nature itself.

The secondary problem is that the Megans of the world are few and far between, and I doubt seriously that the United States Army could raise a single combat ready, all-female company of Megans that could stand a day-long fire-fight against a company of drunk forty-something Russian reservists.

But Cody is a different story. We have millions of Codys.

And in a national emergency our regular armed forces via conscription could cuss, beat, and otherwise hammer these new suburban half-men into something like fighting men or at least reliable cannon fodder.

Should we ever have to call up our Pillsbury Doughboys and the horde of semi-dysfunctional products of the modern illegitimacy and divorce culture and subject them to martial discipline, they would need to emerge swaggering, believing that they alone are capable of saving the day.

Thus we get to the primary problem. By allowing those rare Megans who possess the strength of an average man and the fighting spirit of Xena the Warrior Princess to stand upon their rights so that they may express their patriotism, or indulge their masculine side, or live a fantasy, we strike at the self-confidence of the legions of man-boys that our feminized, post-agricultural society has produced and that we may one day need.

"If she can do this, anyone can. I'm not special. I'm not needed."

But the truth is that she cannot do it. He is special because he can be made to do it, and he—as a fighting man—might be desperately needed, and we must have him sufficiently motivated to sacrifice his life in battle if called upon.

It is woman that motivates him thus. She is indeed the foundation of man's morale, for the Codys of America as well as the Joe-Billys. For a man a good, strong, sweet woman is a jetpack for all of his labors and a hammock for his rest. She cannot change him at his core, but she will improve him in every way. On the other hand a joyless shrew will suck the life out of Conan the Barbarian and wipe the smile off the face of Ronald McDonald.

But there is another face of this motivational issue. The presence of the bazooka totin' Megans of America can only decrease the morale and unit cohesion of our combat units, but they will have the opposite effect on the morale of our enemies. As every battlefield commander has known since tribes squared off with clubs, the enemy is often not defeated in a slow grind that kills the last man fighting, but rather suddenly, when the enemy loses its collective will to resist and breaks in flight or surrenders en masse. The *Captain* works hard to bring his enemy to this breaking point. The presence of women in our combat units will not inspire our enemies to surrender but rather will only strengthen their resolve to fight on to avoid adding humiliation to shame.

The presence of a woman can have a profoundly positive influence on fighting men, but not a grenade chucking one. Putting women in our combat units is needlessly provocative.

◆ ◆ ◆

Since the beginning of time woman has materially contributed to all of man's works, including the defense of her *people.* But in the realm of combat her contribution must be limited to the part that nature has assigned her, a part that is critical for victory, namely sustaining the morale of her man.

This is Human Nature 101. A woman sustains a fighting man's morale, first, by his knowing that she needs him and, second, by her encouragement. The inborn imperative to protect their women and the effect their admiration and support has on them can keep fighting men on their feet in the face of extreme hardship and inspire them to insane feats of courage. In war, where the stakes are life or death, freedom or slavery, that endurance and those deeds may be the difference between victory and defeat.

If Megan actually wanted to help, she along with half a dozen of her girlfriends would sit on the grassy knoll overlooking the pond, and when Cody came up out of the water after being turned end over end by Joe-Billy, they would cheer and chant his name and assure him that next time he would prevail. Then watch Cody pop out of that water and get back in line with a prayer on his lips, "Lord, please don't let the girls leave until I've had one more chance and please, anyone but Joe-Billy! Amen."

XVI. WOMAN THE MOTIVATOR

RUTH WAS THE YOUNGEST child and only daughter among nine brothers, all sired by a bear of a logger and thrown by his bear of a wife, both of them prone to fits of fury, especially when they were drunk, when even the wind lay quiet lest they be provoked. Ruth grew up in what used to be a double-wide, now strapped to the ground with come-alongs and half reclaimed by nature, resembling a burn pile as much as a home. Hardened by disorder and violence, and made cunning by want, she determined early that she would survive come what may, fearing neither man, beast, or tornado. She grew into a tall, big-boned, buxom girl with hair as long and fine as a wild mare's tail and a face as pretty and soft as basswood.

One day Ruth, with spear in hand, was on a long trek through the Great Woods looking for an animal to kill, cook on a sunbaked rock, and eat with her bare hands.

Suddenly a faint, confused noise in the distance grew to reveal a herd of screaming couch potatoes crashing through the underbrush, oblivious to all except their own fear. They passed as quickly as they had come, leaving only the growing sound of the terror they were fleeing.

"Uh-oh!" she said aloud as she spied deep in the woods the lumbering form of Beau bin Polyphemus, the great twenty-foot Cyclops into whose woods she must have strayed.

Taking the only path of escape, she ran after the couch potatoes.

Fortunately she soon came upon a great cave where she sought safety and was not surprised to find the couch potatoes huddled and trembling inside. They were a disgusting sight. A gaggle of pasty-faced, dim-eyed, snack-fed, video-warriors who had just been fired from Walmart for being useless.

A quick survey of the cave's contents only heightened Ruth's anxiety, for it gave every appearance of being inhabited. Besides primitive implements of every kind, upon a hewn shelf were rows of thirty-gallon trash cans, and in the center of the great room its only piece of furniture, the bottom half of a local water tower's tank that had been carried off in a tornado some years back.

The silhouette of Beau the Cyclops in the entrance of the cave, suddenly snatching up one of the couch potatoes and bashing his brains against the roof, then eating him whole, cleared her head. Beau had not been chasing them, he had been herding them! They had not run into shelter, but rather a corral!

The video warriors as one turned their expectant gaze to a prominent boulder and waited for their friend to re-spawn. When he did not and the maniacal fiddling of their air joysticks produced no effect whatsoever, all pandemonium broke out.

When Beau began stuffing large branches under the great water tower basin, the horrifying reality came to Ruth. He meant to can them all! Those weren't thirty-gallon trash cans on the shelf; they were thirty-gallon Mason jars! Through the dirty glass she made out twisted human forms, and closer inspection revealed that they had not even been field dressed. "Now that's just nasty!" she said.

In all her miserable childhood, she had never imagined coming to this end. But anger rather than despair swelled up within her ample bosom, and with spear in hand she made as to charge the presumptuous herdsman. But better sense prevailed: She must make the cowering couch potatoes fight.

She would ascend the boulder, quote Braveheart's speech and with raised spear cry out, "Do you want to live forever?!" and charge Beau the Cyclops—doubtless not followed by even a one of this pack of rabbits. "No, that won't do," she said to herself as smoke began to crawl up the sides of the water tank.

Then from her innermost being another idea came to her, and swallowing the pride of a Caesar she cried, "Save me!" At this the spuds ceased their wailing and turned their glistening eyes to her. "Save me, you towering sons of Conan!"

Keeping their attention in a headlock, she darted from spud to spud. She praised them as warriors, cursed them as cowards, swooned in their arms, laid her head on their shoulders, batted her eyelashes, cried a river of tears, and she struck that flint with her wiles until it finally let off those sparks that finally set alight that little piece of dry tinder in their shrunken souls.

Then the spuds, before her eyes, morphed into something like men.

Then one of them did rise and give Braveheart's speech, and he did lead a charge, and the former couch potatoes did follow him. And together they threw themselves upon Beau bin Polyphemus with all the latent fury of their sex and they smote Beau down dead at their feet.

Though not before Beau had killed most of them and rather gruesomely, but that's beside the point.

But the survivors told the tale and a legend was born, and Ruth accepted the hand of the new Braveheart who worshipped her all the days of his life.

The End

The moral of this story is not that there are never enough fighting-Ruths to kill a cyclops, though there never are. Neither is it that woman can manipulate man to do her bidding, though this is certainly true. Nor is it that woman admires brave men and despises cowards or that man is hardwired to follow a man into battle and not a woman, though this too is true.

No, the moral is that nothing motivates a man like a woman's need or encouragement. The tears and cheers of a damsel in distress are jet fuel for what innate courage he has. This is true of couch potatoes and doubly true of real men.

◆ ◆ ◆

It is ironic to me that the least important argument for not putting women in combat is given the most attention in the media by proponents and opponents alike, namely man's extreme advantage over woman in physical strength and endurance. However understandable the preoccupation with the question of *man's vs. woman's strength* (and there is no getting around man's advantage), the truth is that if God (or *Accident* if you prefer) had granted woman equal strength and endurance with man it would only put a dent in the arguments against institutionalizing women in combat. Combat would still be an extension of man's unique nature and antithetical to woman's.

But before I get into the chip shot that is physical strength and endurance, I want to give the most important reason why we must not institutionalize women in combat.

It's wrong.

XVII. THE IMMORALITY AND THE DEGRADATION

A CAREER, SMALL-TOWN COP once told me the worst beating he ever took in the line of duty was at the hands of an 80-year-old woman suffering from dementia who had become uncontrollable and a danger to herself. It had been decided that she must be carried to the local hospital, an idea that she resisted like a feral cat being stuffed in a bag, and it fell to his lot to restrain the frail creature and ride with her in the backseat of a squad car. Violent as she was, he would not subject the elderly woman to the degradation of being cuffed.

"I couldn't hit her. I couldn't mace her. I couldn't put her in a choke hold. All I could do was take it. You know, we're told our whole lives that you don't hit a girl."

Noble sentiment indeed and a rung in the ladder that ascends to higher civilization, and certainly part of the inheritance of the Christian West.

You don't hit a girl!

Not that those men who have embraced the faith over the last two thousand years immediately ceased to beat their wives, but uncountable and unknown except to God are the hands of husbands stayed over the centuries because of Christ's teachings on the treatment of the weak and defenseless and those teachings have sunk deeply into our collective consciousness as Westerners.

In one of the great ironies of modern times, it is *Feminism* that has come closest to yanking this noble sentiment out of our culture (much like a healthy tooth), an unintended consequence of its faith, which makes it par for the course for *Progressivism*.

It is bold-faced hypocrisy that allows *Feminism* to play the role of the defender of woman, while at the same time—as a consequence of its war on human nature—the status of woman has been degraded in almost every way.

For example, for decades *Feminism* has successfully fought against the sexual harassment of women in the workplace, though to be fair, with the approval and support of good conservative Christians. Fine, we shouldn't smack our secretaries on the behind and call them "Toots."

Yet at the same time *Feminism* has achieved a *Sexual Liberation* for woman that has degraded her to a sexual object in popular culture and made chastity synonymous with prudery.

For decades the United States' Propaganda Ministry, aka Hollywood, encouraged by *Feminism* has systematically debased women sexually and promoted the basest of them as positive role models who have tempted masses of American girls to reject the old ethos of sexual self-restraint and indulge in the new ethos of self-indulgence.

Feminism has even gone so far as to promote prostitution as a legitimate occupation: whores transformed into "sex workers." To be clear here, *Feminism* has produced a divorce and illegitimacy culture that has produced masses of semi- to fully-dysfunctional young women. The most sought after prostitutes have always been and will always be young women. Just as little girls are vastly more naive and vulnerable than young women, young women are vastly more naive and vulnerable than thirty-year-old women. *Feminism* would tempt these semi- and fully-dysfunctional young women, who by dint of their poor upbringing, lack of mature and wise protectors, and the natural immaturity of their age, to enter the one profession that all of history, common sense, and an ocean of anecdotal evidence screams will crush their souls, quite possibly wreck their bodies, and absolutely make them despised.

The problem is that there is something called *respectable* and *disreputable* and they are everywhere recognized because they are born of our DNA. Bravery in man and modesty in woman will forever and everywhere be admired. Likewise cowardice in man and promiscuity in woman will forever and everywhere be despised. These are unbreakable laws of human nature, and propaganda, no matter how sustained or relentless, will ever change or redefine them. It can however wreck lives and cultures.

Like masculine bravery, feminine chastity is respectable because it leads to the good. Feminine promiscuity is disreputable because it leads to the bad. It's just the nature of things.

The result of *Feminism's* successful war to liberate woman from the shackles of feminine decency and monogamous marriage has not led to her fulfillment or equality with man, but rather to the degradation of masses of women, to their self-loathing, and to man's contempt.

But with women in combat *Feminism* may have topped even this staggering achievement. While with one hand *Feminism* has given cultural and legal protection to wives against physically abusive husbands, with the other hand it has now pushed woman literally into the crosshairs of our nation's battlefields.

We're not supposed to hit a girl or beat our wives, but it is suddenly OK for our enemies to hit our women as long as they are in uniform, in which case they can shoot her legs off, run a bayonet through her uterus, or brain her with the butt of their rifles.

But for *Feminism* it's all about equality, so it must want our soldiers to be able to maim and kill our enemies' women as well.

Then what about female POWs? It will forever remain the fate of male POWs to suffer extreme neglect, brutality, summary execution, and mass slaughter at the hands of their captors. Likewise rape has always and will forever remain the fate of women who fall into the hands of their *people's* enemy. It is the black side of war, and *Captains* in history have a dismal record of preventing it.

American female POWs will be raped—likely at the point of capture in the heat of the moment, but certainly in their place of incarceration. The temptation will be too great for their captors and, quite frankly, concentrations of the vulnerable have always been creep magnets. If nothing else she'll submit to survive.

But it is not just woman that is degraded. It is a natural and a healthy sign for the men of any given *people* to have a particularly visceral reaction to their women being abused by the men of another *people*.

But if we make women equal in combat and subject them to the violence of other men, then her fellow male soldiers must crush their natural inclination to protect one of their own women. They must altogether cease to view her as a woman at all, but rather as a man and deny her that special status that man is naturally inclined to grant to the women of his own tribe. This inclination is born of his DNA, which first germinates as he lies in his mother's arms, blossoms in the embrace of his woman, and comes to fruition when he holds his first baby daughter. For man to view woman indifferently is exactly what *Feminism* wants. *Equality* is the supreme good even if it means making both man and woman less than they were before. Thus dies not only a healthy natural inclination to protect his own, but the elevated Western idea of the lady and the gentleman.

Though the idea of the *gentleman* is not as strong in the West as it once was, it still rings true that a gentleman does not abuse a lady. (Likewise, a lady never gives him cause.) Yet it is absolutely necessary for the *Captain* to subject his men to physical and emotional abuse, alternated with encouragement, in order to harden them for battle so they do not fold like rag dolls in the first shock. That is how you train men for battle as well as football. You build them up, break them down, and then repeat. It's good for them, and the young football player or army recruit, as a man, instinctively understands this. But woman is not built up by stress and abuse the way man is but rather broken down by it. Additionally, the nature of combat itself requires the *Captain* at times to present the severest disposition towards his men to maintain discipline and strengthen their resolve. If necessary they must fear him more than they fear the enemy. But a gentleman does not treat a lady thus. In combat itself, the *Captain* must at times literally sacrifice the lives of his men. The *Captain* of a female soldier may not place a higher value on her but rather must lower it, so that it may be the equal of his men.

But for a man, much less a gentleman, to intentionally send a woman of his tribe to die for him is as contrary to his nature as a man as it would be for a woman to send her own child to die in her place. To do so he, or she, must crush an innate instinct that has kept the race alive since the dawn of human time.

Indeed, the *Captain* of a female combat soldier cannot be a gentleman if he is to do his job, and her fellow soldiers cannot be her brothers, because a brother never allows his sister to be degraded. Brothers understand this. Sisters appreciate the sentiment.

By degrading her, he degrades himself.

Furthermore, a woman's tears are her God-given (or *Accidents* accidently given if you prefer) means of releasing and coping with her pain and fears, and they will have a naturally powerful effect upon any halfway-functional man, even if he's cut from coarse cloth. But her *Captain* and fellow soldiers must ignore them, and she must suppress them, lest she be perceived as anything but a man.

Then there is the nature of military life itself. Historically, armies in the field have always been environments of physical violence and sexual license. This is not surprising considering the high concentration of young unmarried men (or men without their wives), regardless of their class, in a semi-lawless environment. We hope that our officers and senior NCO's are indeed "gentlemen" who will keep the men that fill their ranks in order, but their priority is to mind their combat effectiveness in the face of an enemy that is trying to destroy them, not their P's and Q's.

Camp life in peace time is by nature rough and unsafe for soft men. It is certainly not a healthy environment for women, which explains to some degree the cases of sexual abuse in today's military—yet another unintended consequence of feminism's disregard of nature, and you can bet your bottom dollar that *Feminism* will not take responsibility for this problem that it has created by insisting that men and women be integrated in the one human arena where nature and decency demand their separation.

Lastly, woman possesses an inborn imperative to protect her children, and even halfway functional women can give up alcohol, cigarettes, and other addictive drugs during the term of their pregnancy to protect the health of their unborn children. If it is wrong for *We the People* to intentionally put our women in harm's way, then surely it is even more profoundly wrong to risk the lives of unborn children. Current U.S. military regulations require women to report pregnancies, at which point they are given medical leave. But of course it is typically at least a month after conception that a woman becomes aware of her pregnancy. If she happens to be deployed in a combat zone, or even in a dangerous support position, then the child she is carrying unawares remains highly vulnerable. But that is not the only concern. When she does become aware of her pregnancy, can we rest assured that she will promptly report it? Isn't it possible, even likely in many instances, that under pressure to prove herself the "equal of any man" and out of a sense of obligation to her unit she will conceal her pregnancy for as long as she can, exposing—even against her better instincts—her child to a more prolonged period of danger?

♦ ♦ ♦

Men treating women with respect is a sign of their *healthy-nature*. Being worthy of it is a sign of hers. Integrating our combat units can only degrade her and him. But putting her in harm's way is immoral and should only be allowed when the last man lies dead.

XVIII. AN ILLUSORY BALANCE OF POWER
No Technological Fix

IN ORDER TO MAKE WOMEN in combat believable the media has promoted two dangerous illusions: first, that physical strength and endurance in combat are no longer decisive factors and, second, that woman can close the gap with man. Both ideas are equally absurd.

How do you gauge the maximum strength of a man or woman? Imagine someone much stronger than you grabbing you, smacking you around, dragging you across the yard, and trying to drown you in a watering trough. No matter what your age or sex, when someone is trying to drown you in a watering trough, you instinctively call up every last reserve of strength you possess in order to prevent it.

The same rule applies to armies locked in battle. When two *Captains* face off in a contest where death and defeat are the rewards of failure, each one will call upon all of his resources, not limited to but certainly including the physical strength and endurance of his soldiers, and strain them to the breaking point in order to avoid such a fate, to say nothing of attaining sweet victory.

Like two wrestlers facing off, each *Captain* attempts to put his strong side against his adversary's weak side, to attack when least expected and his opponent least prepared. He attempts to deceive at all times in every way. Both *Captains* strive to deny their opponent rest and comfort and to inflict maximum pain and suffering while minimizing that of their own men.

For the soldier in combat what this has always meant in the past, especially for the side that is losing, is extreme physical and mental strain, whether that soldier is hacking away at a wall of shields, digging entrenchments in the face of an enemy, or just struggling to function on a few hours of sleep per day for days on end. The normal daily cycle of sleeping, working, eating, and resting is replaced with no cycle at all, a chaotic period of toil, stress, discomfort, and pain, with food and rest taken sporadically. As an army is being destroyed, its weakest elements will be burned off. When an army *is* destroyed, its *Captain* and remnants will not be found well-rested and fit, but exhausted, mentally and physically broken.

Innumerable factors will converge to decide the outcome of the contest, but there is one that is universal and timeless and that all *Captains* have some control over—preparation. For combat, part of the *Captain's* preparation has always been the selection and conditioning of his men. Obviously, the *Captain* whose men are the strongest, can endure the most punishment and stay awake the longest, has an advantage. It stands to reason that before ever going into battle he will burn off any weakness in his unit if he possibly can and choose not to carry weakness into battle.

This is obvious, and it is why middle-aged men (to say nothing of old men) are not sought, in spite of their experience, maturity, relative good health, and even martial spirit. And if a *Captain* must, for this or that reason, fill his combat units with men over forty, he can be assured that his opponent will see those units as the weak link and look for the opportunity to throw young crack troops against them.

I will wager that all men between forty and sixty will agree with the assertion that their age group would be a liability for the *Captain* because of their naturally declining physical condition. I will further posit that these middle-aged men are still much stronger and hardier than all but the strongest women between twenty and twenty-five, and that these strongest of strong women are much weaker than the average twenty-five-year-old, male combat soldier.

Furthermore, should a power that is the United States' equal ever rise and try us in battle, we can rest assured that the combat units that it throws against us will not consist of its average men, but rather its best and strongest. That would be the smart thing to do at least.

It is a manifest truth that should not need to be pointed out that woman's physical strength and endurance cannot compare to man's. Any woman that has been beaten by a man or just in the passionate grip of her lover can testify to this. He is simply far less likely to be drowned in the aforementioned watering trough than she.

He'll keep trudging through the snow carrying twice the load when she collapses. A typical 150-pound, well-conditioned young man can lift a prostrate, wounded 200-pound comrade in full gear onto his shoulders and, with lungs burning and heart pounding, carry him to safety. It is the rare woman that could. Once conditioned, a man can march all day, trench until dusk, take his turn on guard duty, and do it day after day. In a forced march his female counterpart will be burned off like slag.

Our physical differences are an unbreakable and unchanging law of human nature, and virtually every female soldier that supposedly passes the physical tests of our combat units is only proof that the standards have been lowered or the tests gamed.

Furthermore, every fourteen-year-old girl that beats a fourteen-year-old boy in some physical contest or every woman that excels in strength or endurance in some way comparable to man amounts to zero evidence that evolution is speeding up and the strength gap is being closed.

If we insist upon filling our combat units with, say, 10% to 20% women who on average possess the physical strength of weak men (or, at best, of average men), the only way that it will not put us at a disadvantage is for our *Captain* to make physical strength much less relevant than it has always been. That is, he would have to draw upon technological innovation to offset the disadvantage.

Proponents of women in combat are banking on this, namely, that by means of technological advancements the disparity between male and female body strength will be rendered irrelevant.

Let's say that our *Captain* via technological innovation manages to reduce the weight of his troops' gear by 50% so that his women soldiers can now carry the gear that he needs them to in order to achieve parity or superiority over their enemy male counterparts. The problem is that his opponent will eventually acquire the same lighter gear, and he'll make his men carry just as much as they always have, which means they will carry twice the ammo, rations, armor, and gadgets, all of which will give him an additional advantage.

If our *Captain* has the technological edge that allows him to send the pain over distance to keep his opponent bleeding and exhausted—let's say by way of infrared-guided javelins—and the technology to prevent his enemy from returning in kind—let's say infrared-guided javelin jamming devices—rest assured that our enemy will eventually acquire the means to do the same to our *Captain*.

History is clear on this point. Technological advances diffuse quickly and are never monopolized for long.

As long as our *Captain* can maintain circumstances that are within the strength and endurance range of his weak men and female soldiers, then no problem, for then women can endure what for the hour passes for combat. This is what is happening now.

The problem is that his opponent is pushing his men and material to the absolute limit and moving Heaven and Earth to control said circumstances to his own advantage and to the disadvantage of our *Captain*. If history has taught our *Captain* anything it is that he should at all times expect to lose control of the situation and be forced to strain his all to turn the tables, reestablish equilibrium, or at least survive.

This straining of his all will forever include taxing to the extreme the endurance of his men, even those whose job it is to stare at computer screens or struggle to get materiel moved from point A to point B.

It stands to reason that in his preparations our *Captain* will not only equip his soldiers as best he can, train them relentlessly in the tactics that he believes will be effective, and feed and doctor them to keep them healthy, but that he will also pick the strongest and toughest and exercise them mercilessly so that they are as strong, tough, and interchangeable as they can be, no matter what job they have—even if it is pushing paper or driving a truck—just in case.

Just in case it hits the fan. And in war, it always has, and will forever hit the fan.

In combat men will forever come face to face with each other or be required to do the hard thing with their muscles, or to endure extreme pain, hunger, and sleep deprivation. Our enemies will always see to it, and we'll see to it that they have to do the same.

And in spite of our momentary military supremacy, Afghanistan and Iraq have not been exceptions.

When we are fighting in space suits in zero gravity, strength and endurance will remain, at the end of that cold eternal night, when the last drop of blood spins frozen, a decisive factor in combat.

XIX. THE PURPOSE OF THE UNITED STATES MILLITARY

THE PURPOSE OF ANY MARTIAL group is an extension of the will of its leader. If he commands the obedience of his soldiers, and if together they possess the discipline, arms, tactics, and strategies that enable them to force their will upon others, then he will achieve his purpose.

In human affairs violence is the ultimate expression and arbiter of will.

The powerless are always at the mercy of the powerful, and if they are to live in peace, much less prosper, they are dependent upon the will, and thus the values, of the powerful.

These are unbreakable laws of human existence.

One of American culture's greatest strengths is the submission of armed power to the will of our constitutional, republican, civilian-led governments, whether the armed power of the local sheriff or the United States Marine Corps. This is a value that permeates the officer corps in the United States military, and God (or *Accident* if you prefer) be praised. We have simply never produced anyone like a Caesar or Oliver Cromwell.

The collective values of any given military are of supreme importance as they determine whether or not that military will perform the role of protector of the *people* or turn upon them as oppressor, as since the beginning of time so many militaries have. In the US military it is the values of our commissioned and noncommissioned officers that keep our humble giant on its leash and held in such high esteem in the eyes of the public it protects. Honor is its chief paymaster, not plunder and rapine.

Values are supremely important if our military is to be kept on its leash, but this leash must not become leg irons. The values of patriotism, courage, duty, honor, wisdom, intelligence, deliberation, discipline, yes indeed! But there is no role within the military for democracy or republicanism, or fine notions of fairness. The military by its very nature must be a dictatorship, a world of command and obedience.

If our military is to succeed in its mission as our protector it must remain immune to values that weaken it, including the swirling cultural hurricanes and shifting moral sands that mark our times.

It is not after all the purpose of our military to *reflect* what our society is. Its purpose rather is to *protect* what our society is.

Its purpose is not to promote the agenda of party or faction, regardless of its ostensible good. Neither is its purpose to serve its own members, or support industry.

Its purpose is not to reward the heroic, honor the fallen, care for its wounded, or pay pensions to those that have served faithfully.

Its purpose is not to pay for college, or to give vocational training, travel opportunities, or adventure. Its purpose is not to sustain its own traditions or to allow sons to continue in the footsteps of fathers and grandfathers.

Its purpose is not to give equal opportunity to all people to be all that they can be or to give advancement opportunities to those that seek promotion.

Its purpose is not to bedeck the uniforms of career officers with medals and ribbons or to be a stepping stone for them en route toward a lucrative second career as a consultant or contractor in the defense industry or to ingratiate themselves with the powers that be.

Its purpose is not to kill with swords, arrows, or bullets, or to fire artillery, sink ships with torpedoes, or drop bombs from planes.

Its purpose is not to provide opportunity to indulge a martial inclination or a sense of patriotism.

Its purpose is not to provide self-centered women an opportunity to prove they are the equal of men or to indulge a butch fantasy.

Its purpose is not to shape the culture that called it into being.

The purpose of our military is to protect at all times the *people* and interests of these United States and the sovereignty of our constitutional republic that we may live according to the fashion that we choose; to reduce the number and power of our enemies by means of the threat or the use of violence; to be able at all times to kill, maim, or make captive our enemies, and reduce or destroy their will and capacity to resist our government.

Every man or woman associated with the military, and every decision they make or action they take, from the mundane to the consequential, is in support of that violent moment when the flesh is torn and the thing is broken. If our military fails in its mission, then what freedom and security we have will be a gift from a power other than our own, a power to whom we must servilely doff the cap. And if there is one thing both *Conservative* and *Progressive* Americans can agree upon, it is that we will doff the cap only out of respect.

Senior officers in our military bear little blame when they obey stupid but lawful orders and can be forgiven when for tactical reasons they momentarily keep their heads down in the face of a *bad idea* that has gripped their civilian masters in government. Firm protest, lawful resistance, and strict obedience to lawful orders must be their motto in such times. It is these values that keep the unstoppable beast that is the United States military both primed and on its leash.

But those senior officers that have actively supported institutionalizing women in combat are fools, or cowards, or climbers, or all three. They are poison in the body, arsonists in the attic. They are aiding and abetting a civilian enemy within the body politic that does not care one whit about the mission of the US military and would see it compromised as long as it reflects their vision of the world.

Stupidity, cowardice, and self-aggrandizement have no place in the senior leadership of the US military. Such men cannot be tarred and feathered and whipped through the streets (because that would be wrong, I guess), but they should be driven out of the military and into corporate America to work in mailrooms and lobby kiosks where they belong.

XX. PARTING SHOTS
A Call for Sanity

THE CULTURAL REVOLUTION of the 1960's was aptly named, and since then it has provided us three clear markers that show us just how deeply it has penetrated our collective mind and just how foolish that collective mind has become. Those markers are abortion, so-called gay marriage, and women-in-combat. All three are manifestly contrary to human nature, irrational, signs of a deep cultural rot, and fraught with negative consequences.

Stating the obvious, combat is a man's world. It is wholly of man, an extension of his particular nature, and its possibility the first necessity after air, water, food, and shelter. It is the ultimate means whereby the questions of who lives and who dies, or who rules and who obeys, are answered. It is the hell on earth that man creates, and since the stakes are life and death, they could not be higher. It is an inferno where the twenty- and thirty-something alpha-males of one tribe fight their counterparts of another, to take from the other or keep what is theirs, straining every fiber of mind and body to rip the throat out of their adversary by any and all means available and necessary.

It is a world of hunger, thirst, exhaustion, and pain. A world of blinding rage and panic, of chest-thumping howls of victory and soul-crushing despair—a mind-grating world where you are hunted by men as desperate, determined, or cruel as yourself.

It has always been this way and will forever be this way because it is born of man's unique human nature that compels him to seek or protect his own interest in a world of limited means, much danger, and unavoidable decay.

What combat is not is toting a machine gun in a combat theater, standing guard, or executing POWs. It is not target practice, pushing buttons, or being decked out like G.I. Joe. It is not hanging out the side door of a helicopter like a movie star, hopping from place to place to kick in a few doors, doing a couple of back flips, snapping a couple of necks, maybe giving a clever monologue, and then hopping back to the base in time for the buffet.

It is not for boys. It is not for old men. It is not for video-game-warrior-beta-males. It is certainly not for women. They could not endure it in the days of swords and horses, and they cannot endure it today in an age of rifles and Humvees.

Yet in the name of *Freedom* and *Equality* and their love-child *Feminism*, we may choose to put women in combat, but by doing so we are mixing iron and clay. Besides weakening combat strength and unit cohesiveness, we will also undermine one of man's most primordial motivations to fight—his love of his woman and home.

It is the rare woman that can endure the rigors of forced marches or face off with a furious 180-pound, twenty-something man. Even in our nation of 300-million could we make a full strength infantry company of such Amazons? That Hollywood constantly parades before us 110-pound beauty queens that easily KO grown men while not smudging their makeup or breaking a sweat does not mean they exist. They are fantasy, every bit as much as the exploits of Wonder Woman.

Yet we ignore reality, and indulge fantasy.

But if we continue down this road, our collective delusion will vanish like a pleasant dream upon waking to find our house being consumed by flames. The United States will not always possess the overwhelming military supremacy that we have enjoyed over the last several decades, a supremacy that has allowed us to daydream thus. The day will come when suddenly, like a bolt from the blue, we will find ourselves grappling with an enemy that is our equal or even our superior, who will throw against us his best, strongest, and bravest men, and if we have alloyed our combat units with women they will pull our britches down and whip our ass bloody. The great danger then will be that the notion that woman is the equal of man in combat will sink so deeply into our military culture that we will not even then be able to recognize the cause, much less correct it, thus paving the way for future tactical or even strategic defeats.

History provides a nearly endless number of examples of the unexpected and shocking in war, but perhaps the Tet Offensive, the

117

Chosin Reservoir, and the Battle of the Bulge are still recent enough to ring a bell in our collective memory. Or perhaps a more ringing example would be the movie *Black Hawk Down* which dramatized the events of 3–4 October 1993 in Mogadishu, Somalia where the US Army suffered a stinging tactical defeat. But it was only one night of hell. Only a snapshot of what can go wrong in war.

But when this happens, as it invariably will, over and over again, we will need our combat units to be chock-full of the same young, tough, fighting men who can withstand the blow, and give us time to regroup, counterattack, and achieve victory, or as at Mogadishu, to at least get most of them out alive, avoid disgrace, and shower the service with honor.

This is what General Barrow was getting at when he said bluntly that *women can't do it!* Yes, they can be in uniform and perform well tasks that soldiers, sailors, airmen, and marines must perform. Yes, women can endure combat when our *Captain* controls the battlefield and thus can limit the strain that his soldiers must endure. But so could old men and fifteen-year-old boys. But as our *Captain* prepares for combat, he is a fool if he does not assume that he will, from time to time, lose control and thus prepare accordingly. It is at these times, when the bodies or nerves of trained alpha males in their prime are being shredded, that *women can't do it!* It is at these times that battles are lost.

Putting women in combat is manifestly the dumbest idea in the history of the world, but more importantly it is grossly immoral. What kind of man would hide behind his wife in the face of a mugger? What kind of people would intentionally put their women in harm's way or subject them to degradation? Indeed, only a morally bankrupt people would do so. Only a people who have been too long insulated from the hard realities of life, and who, in a prosperous age, have, in their arrogance, elevated the irreconcilable notions of *Freedom* and *Equality* to the status of sovereign gods and turned their back on unchanging human nature.

Women in combat is but *Feminism's* latest step, and rest assured that subjecting women to the draft is coming. True daughters of revolution, the broken, bitter, radical women that drive feminism will never stop until they have white-washed every distinction between man and woman, and it matters not to them the cultural wreckage they are leaving behind. They do not care one whit about national security, anymore than they do about the American family, and they cannot seriously believe that putting women in our combat units will improve our fighting capacity. Their goal is only to create an illusion that man and woman are the same so that they can declare to man, "We don't need you! We owe you nothing! We are free!" In their war against nature they will bend, crush, and mash humanity into Frankenstein until they can give God the finger, curse Him, and cry out, "We are not of Adam's rib. We are not helpmeet for man! WE *ARE* MAN!"

EPILOGUE
How *Conservatives* Fail to Conserve

CONSERVATISM LOSES to *Progressivism* again and again in part due to Christian women.

The coming of *Feminism* resonated with many Christian women because like Satan tempting Christ, much of what *Feminism* told them was true: Men can be jerks and they shouldn't be; men shouldn't beat their wives; women should be protected by the law; women should be able to work in the new post-industrial economy, etc.

Indeed, taken at face value, some of *Feminism's* proposals were indeed *good ideas.* But knowing its fundamental principles, its goals, all its works, and the bitter, radical *Progressives* that are driving it should cause Christian women to reject *Feminism* forcefully and look for another source of justification and other means of addressing the timeless conflict between man and woman that is born of their respective human natures.

And of course Christian women need look no further than the faith they already profess. But to speak in the name of both *Feminism* and Christ is to serve two masters, which is what Christian women have done for generations. In so doing they have given *Feminism* the cover and support it has needed to promote a view of man and woman that is not only incorrect and foolish, but evil.

This understandable but misguided alliance between Christian women and *Feminism* is representative of what ails *Conservatism* generally and helps explain why it has failed so miserably to stop *Progressivism* in the last 300 years.

Conservatives lose because we are trying to conserve a house built on sand.

We lose because long ago we ceased to view the Father, Son, and Holy Spirit as the foundation of our civilization and have instead, like *Progressives,* come to worship *The Great Assumptions* of the *Enlightenment.* We lose, or at best chase *Progressivism* in circles, because we are arguing from the same assumptions about what is true, right, or best. Like *Progressives* we worship *Individuality, Freedom,* and *Equality,* though in our own way. We don't worship *Feminism* so much, but we do worship *Nationalism*—that is, we worship America.

Until we dethrone these useful ideals and sentiments and subordinate them to the Father, Son, and Holy Spirit, the true fount and foundation of all, we will continue to lose.

Indeed, within the sphere of humanity the only things worth *conserving* are those which draw us towards our Creator. This is the only true human *progress.*

Rejecting the *Enlightenment* and embracing Christ as our sovereign king and his teachings as best and wisest does not mean that we must abandon reason. Indeed the restrictions that He has placed upon us and the self-restraint that He requires are for our and our fellow man's own good. They are indeed *reasonable.* Neither does embracing Christ's sovereignty preclude viewing ourselves as individuals, or using such notions as freedom, equality and nation in the ordering of our lives.

But we must not make gods of them. To do so is indeed to build upon the sand.

APPENDIX
General Robert Barrow Before
The Senate Armed Services Committee
(1991)

File ID: N/A

File Name: SASC Testimony.VOB Robert Barrow. Youtube.

Tape cutting noise [00:07]

Sen. John Warner: [00:08] We appreciate your patience in being with us this morning here, and sitting through all the hearing this morning. We appreciate your being here.

Gen. Robert Barrow: [00:15] Thank you, Senator.

Sen. John Warner: [00:16] I will join in welcoming this distinguished American. I had the privilege of working with him when I was in the Department of the Navy and I watched him move through the senior ranks of the Marine Corps, which movement was predicated solely on accomplishment and merit capability. And we're fortunate to have him, for many reasons. He was awarded the nation's second highest recognition for valor under combat conditions and you speak from a solid base of experience.

Gen. Robert Barrow: [00:49] You're most kind, Sir. I, for a moment thought I had perhaps been rendered irrelevant with some of this testimony. I'll try to be brief and to the point. This is not, should not be about women's rights, equal opportunity, career assignments for enhancement purposes for selection to higher rank. It is about—most assuredly is about—as has already been pointed out, combat effectiveness, combat readiness, winning the next conflict. And so we're talking about national security.

Those who advocate change have some strange arguments. One of which is that the *de facto* women in combat situation already [exists]. That women have been shot at, that they've heard gunfire, they've been in the area where they could've been hit with missiles. But exposure to danger is not combat. Combat is a lot more than that. It's a lot more than getting shot at, or even getting killed by being shot at. Combat is finding, and closing with, and killing or capturing the enemy, if you're down in the ground combat scheme of things. It's killing. That's what it is. And it's done in an environment that is often as difficult as you can possibly imagine. Extremes of climate, brutality, death, dying; it's *uncivilized.* And women can't do it—nor should they be even thought of as doing it. The requirements for strength and endurance render them unable to do it.

And I may be old fashioned, but I think the very nature of women disqualifies them from doing it. Women give life, sustain life, nurture life, they don't take it. I just cannot imagine why we are engaged in this debate about the possibility even of pushing women down into the ground combat part of our profession.

Most harm that could come, would probably come to what it would do to the men in that kind of situation. I know in some circles it's very popular to ridicule something called male bonding. But it's real. And one has to have experienced it to understand it. It doesn't lend itself to easy explanation. It is cohesiveness. I've heard some of those words this morning. It's cohesiveness. It's mutual respect and admiration. It's one for all and all for one. It's believing as a unit that no one else could do what this unit is being asked to do. Perhaps Shakespeare said it best of all, "We few, we precious few, we band of brothers." So that's what it is, and that would be shattered, that would be destroyed. If you want to make a combat unit ineffective, assign some women to it. It's a destructive proposition.

And the thing that puzzles about this, there's no military requirement for it. There's no military need to put women in combat. We have all the men we need for that kind of thing. Male bonding is not peculiar to people who are down say at the infantry level. There's been a lot of commentary about combat flying. It flourishes there. Fighter squadrons, attack squadrons—those men, if they have nothing else in common, it is the belief that they and they alone are able to do what they've been asked to do. And that's male bonding. And we want to play around with this sort of thing, and perhaps do away with it.

And who would be called on to pick up the rifle and do these things in the ground combat area. The ones who don't want to do it, as well as the ones who are not qualified. The young soldiers and [female] Marines.... They are terrific. I know them well. They serve with great skill. They have a spirit about them. They're crème de la crème. And most of them—I've never met one who wanted to be an infantryman. Who wants them to be an infantryman? The hard-line feminists do, that's who wants them to be an infantryman. They have their agenda. And it doesn't have anything to do with national security.

They want to put our daughters at risk. And the other attendant problems to being in such situations, where you have sexual harassment, fraternization, favoritism, resentment, male backlash, all of these things. [It] would be an insurmountable problem for someone to deal with. Who deals with that? Not some faceless political appointee over there in the Pentagon, but the corporals, and the sergeants, and the lieutenants, and the captains would have to maintain good order and discipline and also fight the war. Doesn't work, doesn't work.

And so that moves me to say this. Please, Congress of the United States, *you* keep this responsibility, *you* draw the line. Don't pass it to DoD, don't pass it to the executive branch, because they come and go. You have some continuity, and you would put it in law. They'd put it in policy, but policy can change at a whim. Now we all believe in civilian control of the military. But sometimes that authority is abusive and coercive, and it's done over there quietly. You do not necessarily know about it. When people in the civilian hierarchy of the Pentagon push on the military (the uniform military) do things not because it's the right thing, but because they can do it; they make them do it.

Now I've heard Senator Cohen this morning say, we should not be (you in the Congress) micromanaging. If you gave this authority back, or gave it to the Secretary of Defense, it wouldn't be but a short couple of years and you would be very much into micromanaging. "Mr. Secretary, what is this we hear about that you're going to put women in the Navy

Seals teams?" or whatever it might be. They change the policy to fit the pressure. And you've got it. And so the best place for it to be, the line should be determined by you, and the law should be established by you. And I don't think there's a lot of mystery about where the line should be, it may be exactly where it is now...part by law and part by policy. And if you look at it, and there's any doubts about it, move it back, don't move it forward.

And I do worry about this thing called the draft. I don't know if CSPAN is picking this up or not. I hope they are. They've heard that enough today. It ought to excite the parents a little bit. Maybe you'll get more mail, Senator. I hope you do. I believe that if this thing persists, that's a logical conclusion, that somewhere down the road, women would not only register for it, but if we had a draft they would be compelled to serve, and yes, they could end up in the infantry.

I reflected last night at some length about places I've been, things I've done in my forty-one years in the Marine Corps in three wars.

Sen. John Warner: [10:47] And I think General you should put aside modesty and put in the record today the exact combat situations and the period that you served in them.

Gen. Robert Barrow: [10:58] Well I was in World War II, Korea, and Vietnam. Command in all three. And I found nowhere in my mental exploration *any place* for women to be down in the ground combat elements. Indeed I went so far as to reflect on one campaign in particular, the Chosin Reservoir, forty years ago. December 1950, North Korea. Probably the greatest—one of the greatest epics of all times. 1st Marine Division, confronting eight Chinese divisions spread out over a long thirty-five to forty-mile linear disposition, north-east—north-south. In extreme cold, -25. Winds out of Siberia bringing the wind chill factor down into God knows what. Mountains. Constant attacking—they attacking us, we attacking them. For days, night and day. Death all about. Frostbite, inadequate clothing. I said, "Suppose we had 15% women, 20% women."

My supposing led me to say I wouldn't be here. I guess Kim Il-sung would be taking care of my bones along with everybody else's in North Korea.

I know about my service. If you persist and push this down into the combat area, it would destroy the Marine Corps. Simple as that. Something no enemy has been able to do in over two hundred years. So my recommendation is to make the law clear and unambiguous, if it needs to be made that way. Keep it to you to do. Don't make it a policy. And draw the line where that too is clear and unambiguous. And that's about all I have to say, Sir.

Sen. John Warner: [13:19] Thank you, General.

ABOUT THE AUTHOR

MARK C. ATKINS lives with his wife of twenty-seven years and their six children in Paris, Tennessee. He has been in real estate for thirty-one years. He and has wife founded a small K-12 classical school in 2004.

SHOTWELL
COLUMBIA   So. Car.
EST. 2015
PUBLISHING